SpringerBriefs in Applied Sciences and Technology

T0171972

SpringerBriefs present concise summaries of cutting-edge research and practical applications across a wide spectrum of fields. Featuring compact volumes of 50 to 125 pages, the series covers a range of content from professional to academic.

Typical publications can be:

- A timely report of state-of-the art methods
- An introduction to or a manual for the application of mathematical or computer techniques
- A bridge between new research results, as published in journal articles
- A snapshot of a hot or emerging topic
- An in-depth case study
- A presentation of core concepts that students must understand in order to make independent contributions

SpringerBriefs are characterized by fast, global electronic dissemination, standard publishing contracts, standardized manuscript preparation and formatting guidelines, and expedited production schedules.

On the one hand, **SpringerBriefs in Applied Sciences and Technology** are devoted to the publication of fundamentals and applications within the different classical engineering disciplines as well as in interdisciplinary fields that recently emerged between these areas. On the other hand, as the boundary separating fundamental research and applied technology is more and more dissolving, this series is particularly open to trans-disciplinary topics between fundamental science and engineering.

Indexed by EI-Compendex, SCOPUS and Springerlink.

More information about this series at http://www.springer.com/series/8884

Juraj Ružbarský

Al-Si Alloys Casts by Die Casting

A Case Study

 Springer

Juraj Ružbarský
Design and Monitoring of Technical
Systems
Technical University of Košice
Košice, Slovakia

ISSN 2191-530X   ISSN 2191-5318   (electronic)
SpringerBriefs in Applied Sciences and Technology
ISBN 978-3-030-25149-9   ISBN 978-3-030-25150-5   (eBook)
https://doi.org/10.1007/978-3-030-25150-5

This Springer imprint is published by the registered company Springer Nature Switzerland AG
The registered company address is: Gewerbestrasse 11, 6330 Cham, Switzerland

Introduction

Die casting of metals represents a method of production of accurate casts with the molten metal being injected into a mould at high speed under high pressure. Development of die casting and commencement of utilization of the method dates back to the second half of the nineteenth century. For a long time, the technology of hot-chamber die casting was used. The change occurred during the twenties of the 20th century, then the method of vertical cold-chamber die casting was developed. Ing. Jozef Polák along with a team of employees significantly contributed to the invention of die casting method who at the same time became the first producer of the machines based upon the principle of the vertical cold chamber. The new technology allowed die casting of ferrous and at present of non-ferrous metals with higher melting point including alloys of aluminium, magnesium and copper.

Nowadays, especially the automotive industry offers space for rapid development of progressive casting technologies. Despite high initial costs, the high-volume and massive character of production presupposes fast and reliable returnability of investments. The vehicles consist of a number of components made of steel, cast iron and alloys of non-ferrous metals. The effort is to produce light-weight vehicles which would result in a decrease in fuel consumption. Aluminium alloys being typical for low specific weight, good machinability and castability rank among the most widespread materials from the point of view of foundry industry.

The monograph is focused on the research of the influence of the selected technological parameters of die casting upon the strength and use properties of casts on the basis of Al–Si alloys produced in die casting. Specifically, the aim is to study the relation among the selected technological parameters of die casting with regard to strength and use properties which are represented by the tensile strength and porosity of the casts. As per the research, the individual dependences are assumed between the examined technological parameters of die casting and the properties of the casts on the Al–Si basis. According to the measured results and relevance of the individual technological parameters, the optimal values of the selected technological parameters of die casting shall be proposed to achieve the desired quality of the casts.

Contents

Chapter 1
Al–Si Alloy Casts by Die Casting

1.1 Current State of the Art

1.1.1 Die Casting of Metals

Die casting of metals (Fig. 1.1) represents method of accurate casting, which approaches most to the ideal effort to change basic material into a final product. The die casts are characterized by high accuracy of dimensions, sleek surface, thin walls and very good mechanical properties. In case of die casting , the liquid metal is

Fig. 1.1 Principle of die casting of metals [2]

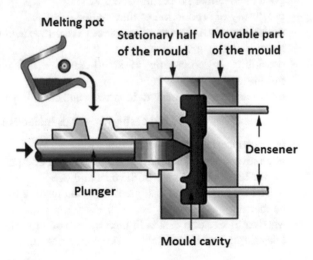

This monograph has been supported by the projects KEGA 006TUKE-4/2017.

J. Ružbarský, *Al-Si Alloys Casts by Die Casting*,
SpringerBriefs in Applied Sciences and Technology,
https://doi.org/10.1007/978-3-030-25150-5_1

pressed into a cavity of a sectional mould at high speed. At speed expressed in units of meters per second, a plunger conveys the melt from a filling chamber by a gating system and through the ingate into the mould cavity. The speed in the ingate reaches the value of several tenths of meters per second. Overall period of the mould cavity filling is rather short, i.e. ones even tenths of milliseconds. Technology of die casting represents a complex of reciprocal relations between alloy characteristics, mould structure and performance of a die casting machine. Basic parameters of die casting include period of mould cavity filling, melt speed in the ingate, mould temperature and metal pressure in the course of cast solidification [1].

1.1.2 Characteristics of Die Casting of Metals

Die casting of metals represents accurate casting which approaches most to the ideal effort to direct change of basic material into final product. Each production technology of components disposes of both advantages and disadvantages. The advantages include the following:

- production of casts falling into low-dimensional tolerance bands, quite often without machining,
- sleek surface of casts,
- possibility of production of components of complex shapes,
- good mechanical properties of casts,
- possibility of production of thin-walled casts,
- utilization of cast intermediate pieces and other metals or non-metallic materials,
- lower material costs,
- possibility of precasting of small-diameter holes with negligible additional machining,
- higher accuracy contrary to identical sand cast alloys.

The disadvantages of die casting of metals include the following:

- High costs related to mould production.
- High investments related to machines and other equipment.
- Lower tensibility in case of die cast alloys.
- A cast cannot be used at higher temperatures due to risk of formation of surface bubbles.
- Maximum size of a cast is limited by size of a machine.
- Die casting requires certain working experience.

From the economic point of view, the die casting is more advantageous when being compared to sand casting as there still exists the option of metal saving (from 10 up to 20%) which is inevitable for cast production. Furthermore, the savings related to production costs in a foundry plant range from 15 up to 30%. Sand casting brings convenience to low-volume casts. In case of medium volumes, convenient is casting into chill moulds and die casting appears to be the most advantageous with high volumes [3].

1.1.3 Basic Methods of Die Casting of Metals

Production of casts employing the method of die casting involves the machines which fall into category of foundry or of die casting machines. Nowadays, the die casting machines are categorized as follows [3]:

1. hot-chamber die casting machines,

    (a) with metal pressing by means of the plunger,
    (b) metal pressing by means of air.

2. cold-chamber die casting machines,

    (a) with vertical pressing system,
    (b) with horizontal pressing system.

1.1.3.1 Hot-Chamber Machines

The hot-chamber machine is typical for its holding furnace with a melting pot which is a part of the machine. The filling chamber and molten alloy are in permanent contact. The chamber is formed in the pot, yet it is separately immersed in the pot, and in any of the cases, the chamber gets narrowed in a gooseneck with a nozzle. Prior to metal pressing, the nozzle is pushed against the gate opening of the mould stationary half. In reverse move of the plunger back to the initial position, the liquid metal flows out of the pot through the hole in the upper part of the chamber. When the plunger moves forward, the hole gets locked. Next, the plunger pushes the metal through the nozzle into the mould cavity. The value of pressure reaches approximately 10 MPa. The plunger is hydraulically controlled. After certain period of time, the metal solidifies in the mould until the cast is produced. Once the mould is opened in forward direction, the cast is removed manually, with tongs or jaws of an automatic removal device. Ejector pins push the cast out of the mould, and consequently, it is removed from the mould manually or by means of the removal device. When opened, the mould must be treated by a spray gun or by a treatment device. During treatment, the lubricating solution is applied onto the active area. After cleaning, the mould is locked and the process of metal pressing is repeated. The instance of hot-chamber die casting based on the principle of metal injection is shown in Figs. 1.2, and 1.3 presents the metal injection by air [3, 4].

Fig. 1.2 Hot-chamber machine with metal being injected by the plunger [3]

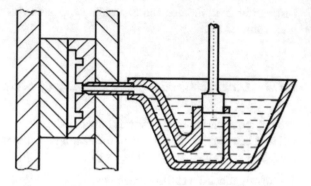

Fig. 1.3 Hot-chamber machine with metal being injected by air [3]

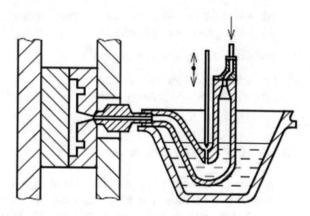

1.1.3.2   Cold-Chamber Machines

Unlike the hot-chamber machines, in case of the cold-chamber ones the holding furnace and the melting pot are separated from the machine. Prior to each pressing, the metal is fed from the pot into the filling chamber manually, i.e. by a ladle or by auxiliary feeding equipment.

In case of the cold-chamber machines, the pressing system includes vertical chamber, nozzles, vertical plunger and lower plunger. To prevent premature flowing of metal into the mould cavity, the lower plunger covers the nozzle orifice. Consequently, the plunger moving downward forces the melt to be pressed into the mould cavity. When the metal solidifies in the cavity, the plunger moves back to the initial position alike the lower plunger the movement of which pushes the melt residues out of the chamber. Alike in case of the hot-chamber machines the mould is consequently opened, the cast is ejected and removed, the mould is treated and locked, and the casting cycle is then repeated. The principle of the cold-chamber machine is shown in Fig. 1.4.

Both the vertical cold-chamber machines and the horizontal cold-chamber machines dispose of a pouring hole. The horizontal chamber with the moving plunger reaches the dividing plane of the mould. During pouring of metal, the plunger reaches

Fig. 1.4 Vertical cold
chamber [3]

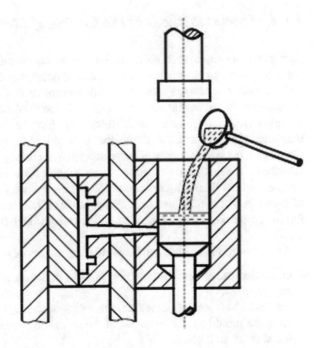

its backwards position and by moving forward the metal is pressed into the mould
cavity. Once the cast is solidified, the mould is opened, the cast is ejected, the mould
is treated and locked, and the cycle is repeated. Figure 1.5 presents the principle of
horizontal cold chamber [3, 4].

Fig. 1.5 Horizontal cold
chamber [3]

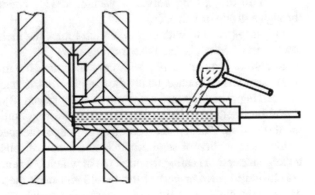

1.1.4 Special Methods of Die Casting of Metals

Except for the basic methods of die casting mentioned afore, further methods have been developed to eliminate negative impacts occurring due to formation of bubbles and pores in the casts. Porosity and microporosity of casts represent the result of influence of speed in the mould and of high speed of melt flowing in the pressure chamber and in the mould ingate. If the speed of filling exceeds the optimal level along with melt viscosity at temperature above the liquid temperature, the turbulent melt flowing occurs which causes disperse mould cavity filling. Certain part of atmosphere is locked in the melt, and once the cast gets solidified, the microporosity occurs. These casts cannot be thermally processed. Shrinking porosity occurs under the influence of high speed of cooling and solidifying. In general, the pores reduce mechanical and fatigue properties, concentrate the tension and diminish the supporting cross section [3, 5].

Gas closure in the melt can be avoided by means of the following:

– controlled movement of the filling plunger in the chamber,
– laminar flow,
– removal of air and of gases from the filling chamber, from the gating system and from the mould cavity by means of the so-called vacuuming,
– increase of melt viscosity.

Elimination of shrinkage porosity can be performed by the following:

– controlled cooling and solidifying,
– action of pressure in the course of entire period of solidification,
– partial solidification outside the mould cavity.

Vacuum die casting utilizes a vacuum device connected to the mould cavity through a group of valves [6].

Direct die casting with crystallization forces the melt into the mould cavity the lower part of which is fixed to a forming press [7].

In case of indirect casting with crystallization, the melt is poured into the tilted filling chamber and consequently the chamber returns back to vertical position. Next, the melt is pushed forward to the mould and pressed [8].

Further method to be mentioned is thixforming both in the locked and opened mould cavities. The method is based on casting of the special aluminium alloy which results in formation of semi-products in a shape of rolls. Prior to casting, the rolls undergo induction heating the outcome of which is a semi-solid state. Next, the rolls are shifted to a lower part of the mould and once the upper part of the mould is pressed, the cavity gets filled with semi-solid metal. The same principle is applied in case of casting in a semi-solid state (thixocasting) [9].

Semi-solid metalworking utilizes fine pellets of magnesium alloy which are loaded into a machine bin. A dosing device feeds the pellets into a screw chamber in which the rotating screw and heating cause semi-solid state of the pellets. Consequently, the magnesium alloy is injected into the mould cavity by a nozzle.

Subfluid-casting is a method in case of which the plunger being cooled by water presses the alloy under the liquid temperature into the mould cavity through multiple ingate [5, 8].

1.1.5   Basic Construction Elements of Die Casting Machines Designed for Metal Die Casting

In general, the machines for metal die casting dispose of identical basic functions. The following rank among the joint characteristic properties:

- horizontal locking mechanism,
- mechanical joint closure predominates,
- horizontal or vertical pressing mechanism,
- hydraulic drive,
- oil or water-glycol is pressure liquid,
- four- or two-column structures.

To achieve desirable quality of casts, the machines must provide safe mould locking, metal pressing into the mould with consequent cast solidification, mould opening, core pulling and ejection of cast from the mould [3].

1.1.5.1   Machine Drive

Die casting machines utilize hydraulic drive. Pressure liquid is mineral oil or heavy inflammable liquid on the basis water-glycol. A pump built in the machine is used for driving the engines. A tank is located in the basic frame of the machine. The most frequently used are the plunger regulating pumps. Additionally, the vane as well as the screw pumps is employed in practice.

In case of a plunger regulating pump, the change of the supplied amount of pressure oil reaches by means of the change of plunger stroke intensity to assure even amount of the supplied oil.

The vane pumps dispose of eccentrically positioned rotor displacing or shifting of which changes eccentricity. When the rotor rotates, the space between vanes changes and thus the sucking or delivery occurs.

The screw pumps dispose of two or three screw spindles in a function of a rotor and in relation to each other those act as locking elements. The thread of a single screw fits into a gap of other screw, and thus, the space between the threads is divided into closed working areas. Silent run and low weight per performance unit are typical for the pumps [3, 10].

1.1.5.2   Locking Mechanism of a Machine

One of the most significant parameters in case of die casting is locking force. To assure size tolerance of casts and to prevent spurting of the liquid metal out of the dividing plane, the mould locking must be ideal. The most frequently used is the frame structure consisting of catch clamps and connecting columns.

Locking force F_u is determined on the basis of opening force F_{ot} striving to open the machine. It is calculated according to the following relation:

$$F_{ot} = p_k \cdot S \tag{1.1}$$

where F_{ot}—opening force [N]

p_k—specific pressure acting upon metal, maximal pressure induced by pressing [Pa]

S—overall area of the cast in the dividing plane [m^2].

The following must be applicable:

$$F_u > F_{ot} \tag{1.2}$$

where F_u—locking force [N].

Recommended is ratio according to the following relation:

$$F_u = (1.1 - 1.2) \cdot F_{ot} \tag{1.3}$$

Opening force F_{ot} must not exceed the value of locking force F_u as the mould could get slightly opened and the metal would leak into the slight opening (primary fin) and enlargement of the area of pressure acting upon metal would lead to enlargement of the slight opening which could result in metal leaking into it again (secondary fin) [3, 11].

Basic types of locking mechanisms include the following:

– hydraulic locking mechanism,
– hydraulic locking mechanism with barrier,
– hydraulic and mechanical locking with barrier,
– mechanical wedge-type locking mechanism,
– hydraulic and mechanical locking mechanism,
– electric locking mechanism,
– hydraulic locking mechanism H,
– hydraulic locking mechanism C.

1.1.5.3 Pressing Mechanism

The main function of pressing mechanism (Fig. 1.6) is to transfer the respective amount of liquid metal into the mould cavity at specific pressure and in the course of the shortest period of time. Speed of plunger can be regulated, and maximal speed exceeds 6 m s^{-1}. The speed is regulated by the hydraulic regulator which allows smooth change of cross section of the feeder piping. At the same time, pressing force can be regulated which might be gradual or continuous. The force can act along the entire length of the plunger stroke or right before its stoppage, which is referred to as resistance pressure [11].

According to structural arrangement and according to method of force regulation, the following can be distinguished:

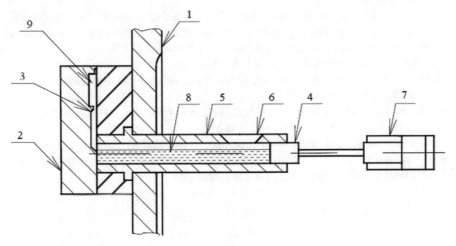

Fig. 1.6 Scheme of pressing mechanism, 1—front strap, 2—mould, 3—ingate, 4—plunger, 5—filling chamber, 6—pouring hole, 7—pressing cylinder, 8—molten metal, 9—mould cavity [3]

- Constant pressing force with zero regulation.
- Gradual force regulation in the course of working stroke of plunger is constant force.
- Gradual force regulation, advanced level is switched on automatically in the course of plunger movement.
- Continuous force regulation, advanced level is switched on automatically in the course of plunger movement.

  Within the scope of production of high-quality products, the pressing mechanism must meet the following requirements:

- Assurance of regulation of speed of plunger in the course of the entire pressing.
- In the first section of the pressing during the plunger movement, the speed should not exceed value of $v = 0.2$ m s^{-1} due to prevention of spurting of liquid metal out of the filling chamber.
- In the second section of the pressing, unless the mould ingate is reached by liquid metal, the plunger should perform accelerated and continuous movement up to maximal speed value which should not exceed $v = 0.9$ m s^{-1}.
- In the third section which should be the shortest as per the machine size (from 15 up to 30 mm), abrupt acceleration occurs up to desired pressing speed—process of mould filling $v = 6$ m s^{-1}.
- Prior to final filling at higher speed, the braking of pressing speed is engaged and speed reaches approximately $v = 1.5$ m s^{-1}.
- Possibility of pressure regulation according to preselected parameters.
- In the final pressing phase, the rapid onset of resistance pressure should range from 0.02 to 0.05 s according to the machine size.

According to the method of structural design and equipment, the pressing mechanisms can be categorized as follows [3]:

– pressing mechanisms with a multiplier,
– pressing mechanism without multiplier.

1.1.6   Machine Auxiliary Equipment

In the process of metal die casting, the main operation of metal injection into the mould cavity is accompanied by a number of additional procedures which are inevitable for a complete production of casts. The procedures include unloading of casts, mould treatment, trimming and additional heating of working fluid. To assure the aforementioned procedures, it is inevitable to employ the equipment described in this chapter.

1.1.6.1   Dosing Device

Manipulators of melt dosing are designed for adjustment of the measured amount of melt of aluminium alloys from the holding furnace to the filling chamber of the die casting machine. The manipulator consists of the key components as follows:

– pedestal,
– drive unit,
– rotary arm,
– supporting arm,
– sensitive elements.

The pedestal structure allows vertical adjustment and rotation of the manipulator. The drive consists of an electric motor and a transmission gear. The output shaft of the screw gear has a rotary arm fixed to it with a chain drive to assure desired position of the supporting arm when the rotary arm rotates. A ladle with a rotation device designed for the ladle rotating is fixed to the supporting arm. The ladle is made of pearlitic grey alloy. The ladle shape is designed to assure only minimal disturbance of oxidation surface when coming into contact with the melt [11].

In utilization of the pneumatic dosing devices, the level of metal in the furnace gradually decreases by means of which the time of the filling chamber dosing is prolonged. To eliminate the drawbacks, it is inevitable to solve regulation of determination of metal level decrease in the furnace on the pneumatic principle.

By analogy based on Ohm's Law for pneumatic system and according to the modified Poiseuille's relation (1.4) for gases, the following can be expressed:

$$Q = \frac{\Delta p}{R_p} \qquad (1.4)$$

where

$$\Delta p = p - p_1 \tag{1.5}$$

Derivation of gas equation of state

$$P \cdot V = M \cdot R \cdot T \tag{1.6}$$

where V—area volume in the pot above the metal level
   M—number of gas moles above the metal level
   R—gas constant
   T—gas temperature above the metal level
can result in the following:

$$Q = \frac{dM}{dt} = \frac{d}{dt}\left(\frac{Vp}{RT}\right) \tag{1.7}$$

Provided that
$V = $ const.
$R_p = $ const.
$T = $ const.
and by substituting into (1.4), the following is the result:

$$\frac{dp}{dt} + \frac{RT}{R_p V}p = -\frac{RT}{R_p V}p_1 \tag{1.8}$$

as $p_1 = 0$, under initial conditions $p = p_0$, $t = 0$, the following solution is reached:

$$p = p_0 \cdot e^{\frac{t}{t_0}} \tag{1.9}$$

in case of which time constant is as follows:

$$t_0 = \frac{R_p V}{RT} \tag{1.10}$$

1.1.6.2   Cast Unloading Device

The process of cast unloading is at times related to operations such as cast testing by means of quality sensors. Freely programmable industrial robots are used which test the unloaded cast by means of sensors.

The most well-known companies dealing with design and production of manipulation equipment is the company of ASEA—Elektrizität in Germany which uses a

Fig. 1.7 Cast unloading machine JOB OT 20 by the Jobs company [12]

flexible system of industrial robots for unloading of casts. Two structural sizes for gripping 6 or 60 kg are at disposal.

The company of WIDO Werkzeug und Gerätebau from Germany supplies the unloading devices for injection weight ranging from 0.1 up to 15 kg. Several programs are at disposal intended for unloading, stacking and inserting tasks.

Another German company of Walter Reis reached in case of robots the area 30 times larger than its inherent area need. Owing to the aforementioned, the space shall be saved for possible arrangement.

The company of Italpresse from Italy supplies automatic unloading machines of IP 21 series for *Al* cast with weight of 10 kg, of IP 22 for the *Al* cast with weight of 30 kg and automatic unloading machines of Estro type for cast unloading and for further shifting to the trimming press.

The company of Jobs in Italy supplies unloading machines JOB OT 20 as a hexa-axial robot controlled by the electric motor with cast weight of up to 50 kg (Fig. 1.7) [12].

1.1.6.3 Mould Treatment Device

Surface and internal purity of casts depends largely on treatment of the active mould part which is provided by the treatment device. The treatment device consists of a reciprocator and a pressure distribution system. The reciprocator is attached to the fixed catch clamp of the die casting machine. It can be mechanically rotated in the area of the dividing plane and by loosening the flange screws. Movement of a treatment block into the mould area is assured by a pneumatic cylinder. The mould treatment is performed by grease and blow nozzles. The air and grease supply into the treatment block is assured by hoses, pipes and a pressure vessel.

In case of the grease circuit, a need of grease jet speed occurs. An equation of relation between the grease jet speed and compressed air pressure in the bin must be derived as the equation of regulation system [11].

The equation of the grease jet in the machine can be written as follows:

$$p = \frac{v^2}{2} \cdot \left(1 + \sum \xi + \lambda \frac{1}{d}\right) \cdot \rho + l\rho \frac{dv}{dt} + l_y \rho g \tag{1.11}$$

where p—compressed air pressure in the grease bin,
  v—grease jet speed,
  $\Sigma\xi$—sum of resistance coefficients,
  λ—pipe resistance coefficient,
  l—pipe length,
  d—inner diameter of the pipe,
  ρ—specific weight of diluted grease,
  g—gravity acceleration,
  $\frac{dv}{dt}$—speed derivation according to time,
  l_y—pipe length in vertical direction.

1.1.6.4   Trimming Press

Control of hydraulic trimming press must be automatically self-guarded. Structural parts such as end limit switches, hand protection devices, and control valves must be driven to achieve self-guarding. Furthermore, a breaking piece assures that the press performs just a single stroke with constantly pushed buttons. In case of trimming press, the safety regulations prescribe the press to prove safety standard by periodical tests which are performed on a yearly basis at least [3, 11].

The company of Weko (Germany) supplies the trimming press with forces of 63, 100, 200, 650 and 1000 kN and with referential numbers WE 480, 650, 1250 AND 1500 and with an electronic control unit and special accessories for cast unloading.

The company of Italpresse (Italy) supplies the trimming press PA 25, PA 35 and PA 50 M with forces of 245, 343 and 490 kN (Fig. 1.8).

1.1.6.5   Device for Auxiliary Heating of Working Fluid

The die casting machine can reach the operating temperature of the working fluid in the course of 3 or 4 h of operation. Auxiliary heating is used when the machine is

Fig. 1.8 Trimming press by the company of Italpresse [12]

switched off for a long period due to repair or planned unavailability time in a factory and when the fluid temperature drops below 20 °C. In case of working temperature drop, the viscosity increases which leads to formation of a film on the filter sleeve once the machine is put into operation again. Thus, signalization of filter failure shall be triggered although the filter has not been damaged. If viscosity changes, the signalization shall turn off. The operator switches on the auxiliary heater before starting the machine. The most advantageous in case of fluid heating utilization is an electric heating body, for instance heating body of TYP 4409 (Fig. 1.9) series consisting of steel heating branch connected to a steel head [11].

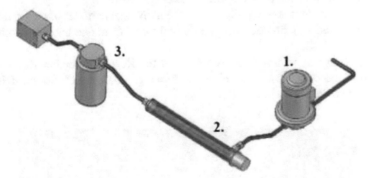

Fig. 1.9 Model of heating body connection, 1—pump, 2—heating body and 3—filter [3]

1.1.7 Process of Die Casting of Metals

The entire process of die casting can be divided into three basic phases [3, 13]:

First phase—prefilling of the filling chamber

Once the chamber is filled with fluid metal, the plunger starts moving. Initially, the plunger moves at rather low speed of approximately 0.2 m s^{-1}. The plunger keeps moving at the speed until it reaches the position behind the pouring hole. Consequently, the evenly accelerated movement commences up to speed of 0.9 m s^{-1}. The speed is controlled by a flow rate regulator. Turbulences of molten metal must be avoided during the first phase as it would cause metal aeration.

Second phase—cavity filling

The mould filling begins once the molten metal reaches the mould ingate. At this point, an abrupt acceleration commences yet the filling is performed at constant speed. The mould filling speed reaches the values of up to 10 m s^{-1}.

Third phase—pressure phase

During the final phase, the melt is forced out with intensive force and fills the entire mould area. Once the filling has been completed, the pressure hold occurs. Pressure intensity in this phase is controlled by means of pressure regulators along with resistance pressure start. The pressure representing the pressing force is significant in the pressure phase from the point of view of shrinkage, formation of shrinks and surface.

1.1.7.1 Metal Flowing

In case of laminar flow (Fig. 1.10), the particles of liquid metal move in the layers parallel to flow direction and reciprocal mixture is avoided.

In case of turbulent flowing (Fig. 1.11), the individual particles of the liquid metal change the speed and direction irregularly and move from one layer to another and get intermixed [15].

The flowing can be described by means of the Reynolds number Re. It is non-dimensional number closely defining the flowing. The characteristics of flowing change and so does the value of Re.

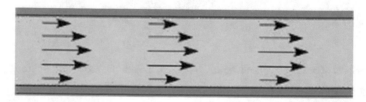

Fig. 1.10 Laminar flowing [14]

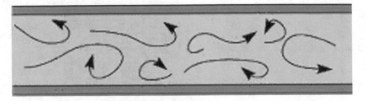

Fig. 1.11 Turbulent flowing [14]

Mathematically, Re can be defined as follows:

$$Re = \frac{v \cdot d}{v_k} \tag{1.12}$$

where v—flowing speed [m s^{-1}],
l—characteristic dimension of piping [m],
v_k—coefficient of kinematic viscosity [m^2 s^{-1}].
Simple definition characterizes Re as ratio between inertia and friction forces acting upon specific part of airflow (liquids). If Re number is high, the forces following from viscosity of the airflow (liquid) act inconsiderably upon the body. Yet should Re number be low, its impact is rather significant.

$$Re = \frac{\text{inertia forces}}{\text{viscosity(friction forces)}} \tag{1.13}$$

The Reynolds number significantly influences arrangement of the marginal layer. In case of low Re only laminar marginal layer is formed. Should Re be critical (Re = 500,000 or 5×10^5), the laminar layer becomes unstable and changes into turbulent marginal layer. If Re is higher critical (over 5×10^5), the mixed marginal layer occurs [16, 17].

From the point of view of front of liquid metal flowing, further types of flowing can be distinguished as follows:

Planar flowing (Fig. 1.12) is characteristic for continuous regular front of liquid metal flowing along entire mould cavity width. At low speed, air bubbles are not closed.

Contrary to mould cavity width, the non-planar flowing (Fig. 1.13) is irregular

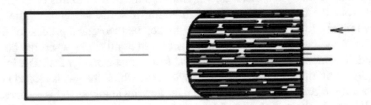

Fig. 1.12 Planar flowing [3]

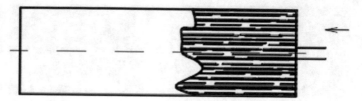

Fig. 1.13 Non-planar flow [3]

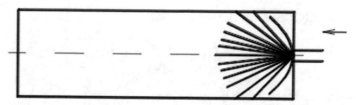

Fig. 1.14 Disperse flowing [3]

within narrower range. At higher speed, air bubbles are closed.

In mould cavity filling disperse flowing (Fig. 1.14) forms a disperse mixture of liquid metal and air.

1.1.8 Mould Cavity Filling

In mould cavity filling by liquid metal, two basic types of filling can be distinguished as follows:

$\frac{S}{S_1} < 0.25$ (mould cavity filling as per Frommer)

$\frac{S}{S_1} > 0.25$ (cavity filling as per Brandt)

where S—cross section of ingate corresponding to cross section of filling jet [m²],

S_1—cross section of the mould cavity to which the gate is led [m²].

According to Frommer's theory, the liquid metal flowing in the mould cavity is typical for the metal flowing through the ingate shall hit the opposite mould wall, which is referred to as hitting phase (Fig. 1.15). Consequently, the metal is dispersed along the vertical walls of the mould cavity and proceeds to the ingate at the place of which is retained by the metal flowing through the ingate. In this case, the mould is filled from the opposite side in the direction towards the ingate. The type of filling is referred to as reverse filling. In practice, the theoretical process of filling is contradicted by the fact that it cannot be achieved at all times when the cooling is increased. The metal hits the mould wall and loses kinetic energy fast. The inhibiting effect may result in formation of metal eddies into which the gas is closed [18, 19].

According to Brandt's theory, the metal jet flowing at low speed is dispersed after having left the ingate and the mould cavity is filled from the ingate towards remote parts of the mould cavity. The process is referred to as gradual filling (Fig. 1.16).

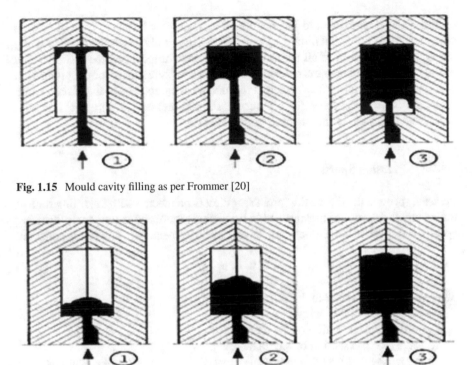

Fig. 1.15 Mould cavity filling as per Frommer [20]

Fig. 1.16 Cavity filling as per Brandt [20]

Diverse approaches of L. Frommer and W. R. Brandt can be reasonably justified. Kinetic energy of the melt flowing into the mould cavity is crucial for the type of filling. According to J. Koch, the following is then applicable:

– If kinetic energy in the ingate is more intensive than metal flowing resistance in the mould cavity, the filling is performed on the basis of Frommer's theory.
– If kinetic energy of the melt in the ingate is less intensive than metal flowing resistance in the mould cavity, the filling is performed on the basis of Brandt's theory.

The mould cavity shape plays an important role. Its geometric arrangement determines whether the flowing metal shall behave according to Frommer's theory, or it shall be carried away to other parts of the mould cavity [18, 19].

1.1.9 Technological Parameters of Die Casting of Metals

The properties and especially the quality of casts are influenced by a number of diverse factors affecting the entire die casting process. The factors include correct mould structure along with the gating system, venting and cooling system of the mould. Other factors should be mentioned as well—for instance, type of casting alloy, quality of produced mould, die casting machine and its attendance and mainly the appropriate setting of all technological and metallurgical parameters. The total of the factors and parameters determines an adequate prerequisite for the production of high-quality die casts. The principal technological parameters of metal die casting include pressing speed, working pressure (resistance pressure), filling time of mould cavity and heat ratio in the die casting process [21].

1.1.9.1 Pressing Speed

Pressing speed in the die casting process is closely connected with the filling time of the mould. It determines metal speed in the gating system and in the ingate. Pressing speed v and speed in the ingate v_1 can be expressed by means of a continuity equation.

$$v \cdot S_p = v_1 \cdot S_f \tag{1.14}$$

where v—pressing speed in the filling chamber [m s^{-1}],
  v_1—speed of metal flowing in the ingate [m s^{-1}],
  S_p—area of cross section of the plunger [m^2],
  S_f—area of cross section of the ingate [m^2].
  The experiment proved that in measurement of speed of the flowing liquid metal in the ingate the actual speed represented from 30 up to 50% of the calculated theoretical speed. Thus, in the mould cavity, the speed values range from 5 up to 15%. The actual speed of metal flowing is influenced by the following [21]:

– alloy viscosity in relation to the temperature,
– loses caused by friction in the gating system and in the mould cavity,
– loses caused by change of jet direction in the gating system and in the mould cavity,
– air pressure in the mould cavity streaming against the metal movement.

The speed of melt flowing in the ingate changes from 0.6 up to 100 m s^{-1} and affects the mechanical properties and internal and superficial quality of the casts. Correct selection of the speed of the mould cavity filling depends on alloy type, on cast complexity, on wall thickness and on ratio between cross-sectional area and cast area. Dependence of the optimal speed in the ingate on the wall thickness and on cast length is shown in Fig. 1.17 [22, 23].

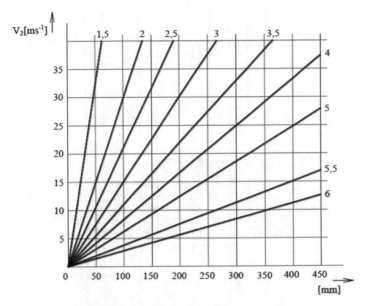

Fig. 1.17 Dependence of alloy speed in the ingate on the wall thickness and on maximal distance of mould cavity from the ingate [24]

The following speed types are distinguished (Table 1.1):

– low speed from 0.6 to 1 m s^{-1} with massive laminar filling,
– medium speed from 1 up to 15 m s^{-1} assuring massive turbulent filling,
– high speed of over 25 m s^{-1} in case of which disperse filling occurs.

1.1.9.2 Working Pressure

Pressure acting upon the liquid metal in the filling chamber should assure the mould filling at suitable speed and during optimal period of time. Hydrostatic pressure must be sufficiently high to exceed the resistance of the solidifying metal mass in the thin cross sections of the mould cavity. Moreover, it should exceed the resistance of gases which remain in the cast. The pressure is transmitted from the plunger through a sprue. The transmission of hydrostatic pressure to the mould cavity is referred to as resistance pressure. To gain the top quality of the cast, it is inevitable to assure the shortest delay period between the mould cavity filling and achievement of maximal pressure. The period is expressed in tenths of milliseconds at time when the cast is not completely solidified. The accuracy of cast moulding and surface roughness depend on kinetic energy of current. The development of hydrodynamic pressure p_H with $S_f/S_F < 0.25$ can be calculated from the relation as follows [4]:

Table 1.1 Speed of metal flowing in the mould ingate for alloys and wall thickness [3]

Group of casts		Zn	Mg	AlO		Cu		Fe	
				Liquid	Pasty	Liquid	Pasty	Liquid	Pasty
Thick-walled 6–10 mm	Simple	30–40	–	0.5–1	2–3	–	2–3	0.5–1	2–3
	Complex	40–50	–	0.5–1	3–6	–	3–5	0.5–1	3–6
Medium thickness 3–6 mm	Simple	40–60	30–40	8–12	5–8	–	5–8	3–6	5–10
	Complex	60–80	30–50	10–20	8–10	8–15	5–8	5–8	5–10
Thin-walled 1.5–3 mm	Simple	80–100	50–60	30–40	–	10–20	–	10–20	–
	Complex	110–120	50–80	40–60	–	–	–	15–30	–
Localized thickness	Simple	80–100	40–60	15–20	10–15	15–20	–	10–20	–
	Complex	100–120	60–80	30–50	–	–	–	–	–

$$p_H = \frac{S_f}{S_F} \rho v_1^2 \qquad (1.15)$$

where p_H—hydrodynamic pressure [Pa],

S_f—cross-sectional area of the ingate [m^2],

S_F—cross-sectional area of the mould cavity [m^2],

v_1—speed of metal flowing in the ingate [m s^{-1}],

ρ—specific weight of alloy [kg m^{-3}].

Development of pressure in mould filling can be divided into three alternatives [3]:

- Value of pressure in the filling chamber acting upon alloy ranges from 2 to 7 MPa at speed in the ingate ranging from 20 to 60 m s^{-1} with thickness of the ingate falling into the interval from 0.3 up to 0.5 mm in case of small casts. In case of medium-size casts, the ingate thickness is of 1 mm and big casts dispose of ingates with thickness of 1.8 mm (Fig. 1.18).
- In casting with hot-chamber machines for thick-walled casts, the pressure acting upon alloy in the filling chamber (Fig. 1.19) is lower and reaches the value of approximately 1 MPa. The resistance pressure reaches the values of 7 MPa and more.
- In case of horizontal cold-chamber machines with resistance pressure, the pressure acting upon the alloy in the filling chamber reaches the value of approximately 50 MPa (Fig. 1.20). The resistance pressure reaches the values of even 600 MPa. The alloy jet speed in the ingate is lower with larger cross sections of the ingate.

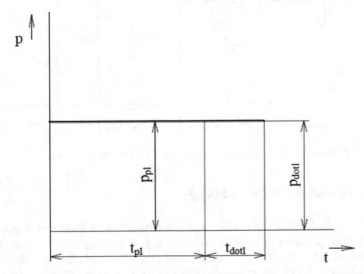

Fig. 1.18 Pressure development in shape-complex and thin-walled casts in case of hot-chamber machine [3]

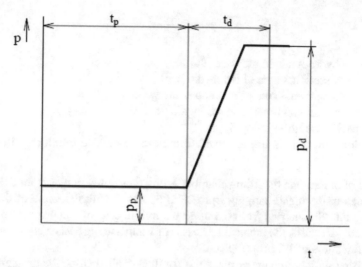

Fig. 1.19 Development of pressure in thick-walled casts in case of hot-chamber machine [3]

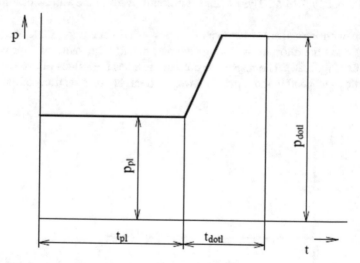

Fig. 1.20 Development of pressure in casting with cold-chamber machines [3]

1.1.9.3   Time of Mould Cavity Filling

Other main technological parameters which in a high-degree influence the quality of the die casts include time of mould cavity filling. It affects especially the cast surface quality and its inner structure. In case of short time of mould filling, the insufficient gas and vapour escape from the mould cavity occur. The gases and vapours get closed in the cast wall, and despite sufficient quality of the surface, the inner structure is disturbed. A long period of mould cavity filling allows escape of gases through the venting system during the movement of the melt front in mould. The inner structure is suitable, yet the temperature at the jet front decreases, and thus, ideal melt blend is avoided and weld lines and sinking occur. It is caused by a risk in case of casts which are dynamically and cyclically stressed. The optimal time should move between long and short period of mould cavity filling and should be shorter than time of cast solidification in the mould. Theoretically, time of the mould cavity filling can be calculated according to the relation as follows [25, 26]:

$$\tau = \frac{V_F}{S_f \cdot v_1} \tag{1.16}$$

where τ—time of mould cavity filling [s],
V_F—mould cavity volume [m^3],
S_f—cross-sectional area of the ingate [m^2],
v_1—speed of metal flowing in the ingate [m s^{-1}].

The onset time, which in comparison with the mould filling should be shorter, can be theoretically expressed according to the following relation:

$$t_1 = \frac{l_0 \frac{S_f}{S_p} - l_1 \sqrt{\frac{S_f}{S_p}} - l_2}{v} \cdot 2.944 \tag{1.17}$$

where t_1—onset time [s],
v—pressing speed in the filling chamber [m s^{-1}],
l_0—overall length of plunger trajectory [m],
l_1—length of melt trajectory in the gating system before the ingate [m],
l_2—length of melt trajectory in the ingate [m],
S_p—cross-sectional area of the plunger [m^2].

In case of constant thermal and physical characteristics, it has been proved in an experiment that the filling time depends solely on wall thickness of the cast. The filling time does not depend on cast dimensions. The optimal mould cavity filling in relation to the wall thickness of the cast (Fig. 1.21) can be expressed by an empirical equation as follows [4]:

$$\tau = 1.6 \times 10^{-2} \times s^{1.984} \tag{1.18}$$

where s—wall thickness of the cast [m].

1.1.9.4 Heat Ratio in Die Casting of Metals

Contrary to other means of casting, the temperature in case of die casting is of higher significance due to stress of the mould. The stress is caused by the thermal shock, and bigger the temperature difference between the temperatures of molten metal and of mould, the more intense is the stress. Therefore, prior to the casting, the mould

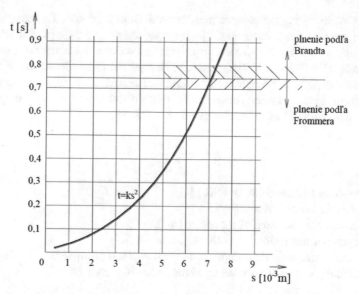

Fig. 1.21 Dependence of optimal filling time of the mould t on wall thickness of the cast s [3]

must be preheated to the optimal temperature. The higher the casting temperature, the higher the mould temperature must be. Alloys intended for die casting usually dispose of the temperature from 10 to 20 °C higher than the temperature at the beginning of crystallization. In order to preserve the inevitable quality of casts, the optimal temperature of each thermal factor must be assured. The thermal factors include the following:

– temperature of casting alloy,
– temperature of die casting chamber,
– temperature of mould.

Higher attention must be paid to casting of the alloy with extremely high temperature into cold mould in case of insufficient surface insulation by the appropriate greasing agent which causes extreme stress of the surface layers of the mould. Furthermore, insufficient temperature of the mould cavity surface results in premature decrease of the alloy temperature which causes occurrence of the weld lines and sinking on the surface of the casts. Inside the casts, considerable inner strain can be observed that is an outcome of undercooling. The undercooling is recognizable for occurrence of the negligible surface cracks [1].

Thermal equilibrium in the casting process can be expressed by the following relation [27]:

$$Q_1 + Q_2 - Q_3 - Q_4 = Q_1 \tag{1.19}$$

where Q_1—amount of heat inevitable for the mould preheating [J],
    Q_2—amount of heat supplied to the casts [J],

Q_3—amount of heat supplied by the cast to the mould [J],
Q_4—amount of heat reduced by cast heat released from the mould [J].
Amount of the heat supplied to the casts:

$$Q_2 = m_k[c_L(t_l - t_s) + l + c_S(t_s - 20)] \tag{1.20}$$

where m_k—cast weight + 0.6 of the gate weight [kg],
c_L—specific heat of the molten alloy [J kg^{-1} K^{-1}],
c_S—specific heat of the solid metal [J kg^{-1} K^{-1}],
t_l—metal temperature in casting [°C],
t_s—temperature of a solid [°C],
l—internal latent heat of the metal [J kg^{-1}].
Amount of heat supplied by the cast to the mould:

$$Q_3 = \alpha F_{odl}(t_l - t_f)\tau_l \tag{1.21}$$

where α—coefficient of heat transfer [W m^{-2} K^{-1}],
F_{odl}—surface of the cast [m^2],
t_f—mould temperature [°C],
τ_l—cooling period [s].
Amount of heat released from the mould along with the cast:

$$Q_4 = m_k c_s(t_2 - 20) \tag{1.22}$$

where t_2—temperature of the cast removed from the mould [°C].
By modification of Eq. (1.19) to the version of $Q_2 - Q_3 - Q_4 = 0$ and by substitution of Eqs. (1.20), (1.21), (1.22) the equilibrium equation shall be as follows:

$$m_k[c_L(t_l - t_s) + l + c_S(t_s - 20)] - m_k c_s(t_2 - 20) = \alpha F_{odl}(t_l - t_f)\tau_l \tag{1.23}$$

1.1.10 Moulds for Die Casting of Metals

The quality of casts produced in die casting is considerably influenced by the mould structure. The die casting mould consists of two basic halves as follows:

- stationary half with the gate belonging to the mould attached to the machine table,
- movable half belonging to the mould attached to the holders; the mechanism for cast release from the mould is included as a standard.

The mould can consist of one or of two cavities (i.e. a single- and a double-cavity mould). In case of a single-cavity mould, a sole cast is produced, and in a multicavity mould, two or more casts can be produced. The structure as well as service life of the die casting mould influence the cast quality and economic efficiency of the production in a high degree. The moulds designed for bog and complex casts are rather costly which is related to labour content and price of the used material. The mould price can be decreased by typification of casts and of moulds according to the dimensions of die casting machines [28, 29].

Die casting moulds are stressed by high temperatures and pressure, by abrupt changes of temperatures and by erosive effect of the melt. The aforementioned factors determine the basic requirements related to mould material:

- chemical resistance to casting alloy,
- preservation of mechanical properties at high temperatures,
- sufficient resistance to thermal shocks,
- good machinability,
- reasonable price.

The steel alloyed with vanadium, chromium, molybdenum, wolfram, cobalt and other elements after heat treatment meets the aforementioned requirements (Table 1.2). To prolong the service life, the mould is chemically treated. Nitriding with diffusive annealing represents the effective mould protection. Nitriding of the mould into depth of approximately 0.3 mm increases the surface hardness with the preservation of the core toughness. The technology of application of protection layers is employed as well. The protection layer of the surface is formed by dipping the mould into a molten borax bath with dissolved compound of vanadium, chromium and niobium. Thus, the reaction between carbide-forming elements and steel carbon occurs and the carbide layer is produced. The bath temperature reaches the values from 1000 up to 1050 °C. Holding time depends on the melt temperature and on desired thickness of the protection layer. To gain the thickness from 5 up to 10 μm, the holding time ranging from 4 up to 8 h is sufficient. Owing to the aforementioned, the moulds are protected against erosive effects of the liquid metal and dispose of high resistance to oxidation and as well as to temperature fluctuations and to mechanical wear. Similar properties can be observed in case of low-carbon steel (referred to as mild steel) alloyed with nitrogen (0.1%) with high chromium content (11%) which is suitable especially for production of moulds designed for die casting of aluminium alloys. Prolongation of service life can be achieved by means of the following precautions:

- regulation of mould temperature,
- utilization of suitable or endothermic grease for the mould,
- utilization of materials for production of moulds with the lowest temperature of expansivity, with the lowest coefficient of elasticity and with the highest thermal conductivity,
- die casting technology with crystallization and processing of semi-solid metal.

The utilization of high-quality steel for production of moulds and surface treatment of moulds is reflected in considerable price increase. In economic analysis, the steel consumption must not be taken into consideration separately yet it must be included into overall costs related to cast production. The analysis can prove that utilization of high-quality steel is economically effective. The mould service life

Table 1.2 Properties of metals and of their alloys used for mould production [3]

Material	Additive element [%]	Modulus of elasticity [MPa]	Coefficient of thermal expansion α [10^{-6} m/(m K)][b]	Thermal conductivity [W m^{-1} K^{-1}][a]	Hardness [HB]	Resistance to melting
Copper	–	105,000	16.5	0.393	60	Excellent
Beryllium bronze	2.5 Be	110,000	16.5	0.0836	400	Excellent
Iron	–	220,000	11.7	0.0752	70	Excellent
Heat treated mild steel	0.4 C	220,000	11.7	0.0627	8	Excellent
Low-alloy and heat treated steel	0.3 C, 1 Cr	215,000	10.5–13.0	0.0418	300–400	Good
High-alloy and heat treated steel	0.3 C, Cr, W, Mo, V	210,000	10.5–13.0	0.0188–0.0334	350–500	Poor
Austenitic steel	0.1 C, 18 Cr	200,000	19.5	0.0209	130	Very poor
Molybdenum	–	350,000	4.9	0.146	170	Excellent
Molybdenum-Titan	–	350,000	3.2	0.125	200	Excellent

[a] At standard temperature
[b] At temperature 20–100 °C

depends especially on type of casting alloy. Prior to casting, the diverse protection or dividing agents are applied onto the mould surface which apart from mould protection facilitates cast removal from the mould. The agents can be solid, vaseline or liquid. The solid agents are grease and wax, and the vaseline ones are mixtures of paraffin, mazut and ceresine, mineral oils, etc. Separating effect is increased by adding the aluminium powder or graphite. Significant drawbacks of solid and vaseline agents are difficulties in automatic application. The liquid agents consist of dissolvent and of additives which together with the dissolvent form a disperse emulsion. In selection of dividing agent, its ability to produce gas should be taken into consideration [30, 31].

1.1.10.1   Gating System

In case of die casting, the gating system (Fig. 1.22) significantly differs from the gating systems of other methods of cast production. It is short and simple. The gating system in case of die casting with the horizontal cold-chamber die casting machines is shorter contrary to the gating system of the vertical cold-chamber die casting machines as the gate does not need to be used.

According to the position with regard to the cast, the gating systems are categorized into the external and internal ones. In case of the internal system, the melt can be directly connected with the cast or through the cast hole, if possible. The external gating system is used more often especially with multicavity mould in case of which they are filled through the ingates and filling of the subordinate sprues is performed once the main sprue has been filled. Other element of the gating system is the ingate with thickness ranging from 1.5 up to 3 mm which depends on the wall thickness of the cast. The point of ingate insertion into the cast is selected on the basis of the character of mould filling and analysis of the cast structure. The correct selection of

Fig. 1.22 Gating system of the die casting mould [18]

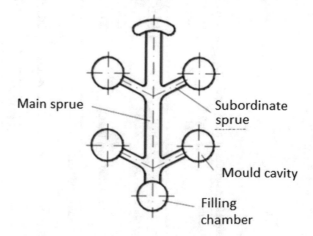

the point of liquid metal pouring into a cast can significantly minimize the occurrence of foundry faults [28, 31].

The suitable point for ingate insertion is selected on the basis of the following characteristics:

- In gradual rejection of the individual venting elements in the course of the mould cavity, filling the alloy jet should be directed to assure the function of the last of the elements even during the final phase of the mould cavity filling.
- To direct the alloy jet so that it would not hit the obstacles such as cores or lugs in the mould cavity.
- To direct the alloy jet in casts with longitudinal openings to assure its parallelism with longer dimension of openings.
- To direct the alloy jet in casts with lugs of rectangular shape to assure its parallelism with longer dimension of openings.

The gating system is designed to avoid metal jet hitting the core. If the jet hit the obstacle, the jet would get split and the direction would change which could result in formation of gas bubbles in the cast wall and intense adherence of alloy onto the core. In case of casts of rectangular shape, it is suitable to place the ingate to the narrower part of the rectangular.

The filling chamber is connected with the ingate by means of a sprue. The single-chamber moulds dispose of a sole sprue with the ingate which is placed in a movable half of the mould. That is due to prevention of disturbance of the front of the standardized filling chamber. To assure almost steady increase of speed of the liquid metal in flowing through the sprue, the diameter of the sprue should evenly decrease along with the distance from the filling chamber towards the ingate. In the last quarter of the distance in the proximity of the ingate, the diameter gets smaller which causes higher acceleration of the alloy jet. Thus, the difference between optimal alloy speed in the ingate and the speed of last part of the sprue becomes smaller as well. Abrupt increase in cross section of the sprue and consequent reduction to the original cross section cause closure of bubbles in the alloy and their carrying into the mould cavity [32, 33].

Alloy speed in the sprue is selected on the basis of the following [3]:

$$v_k = (0.3 - 0.8)v_z \qquad (1.24)$$

where v_k—speed of liquid metal flowing in the main or subordinate sprue [m s^{-1}], v_z—alloy speed in the ingate [m s^{-1}].

Dimension of cross section of the sprue can be expressed by means of the simplified equation of continuity:

$$S_k \cdot v_k = S_p \cdot v_p \qquad (1.25)$$

By substituting Eq. (1.24) into Eq. (1.25), the following relation shall be the result:

$$S_k = \frac{S_p \cdot v_p}{(0.3 - 0.8)v_z} \qquad (1.26)$$

where S_k—cross section of the sprue [m^2],

S_p—plunger area [m^2],

v_p—plunger speed [m s^{-1}].

Coefficient within the range from 0.3 up to 0.5 is specified for the main sprues, and for the subordinate sprues, the coefficient reaches the values from 0.5 up to 0.8.

The coefficient of losses in the main sprue is a sum of the partial factors as follows:

$$k = k_1 + k_2 + k_3 + k_4 + k_5 \qquad (1.27)$$

The coefficient of losses in the subordinate sprue is a sum of the partial factors as follows:

$$K = k_1 + k_2 + k_3 + k_4 + k_5 + k_6 \qquad (1.28)$$

where k_1—influence of the sprue length,

k_2—influence of the wall thickness of the cast,

k_3—influence of the number of the casts,

k_4—influence of irregularity of the thickness of the walls,

k_5—influence of number of the bends of the sprue,

k_6—influence of the cast structuring.

1.1.10.2   Mould Venting

Venting system in die casting serves for drawing off the closed air in the die casting process. In the course of filling the closed air must be forced out by the venting channels through the passage hole. In the mould cavity filling excessive overpressure must be avoided in the mould which causes increase in hydrodynamic pressure of the alloy jet and its separation into secondary jets in the cavity. The venting channels must be of a size of the cross section that must be sufficient for assurance of inevitable permeability of air and gases leaking out of the mould cavity from the entering metal.

Distribution of the venting channels depends on speed of alloy in the ingate and on character of the mould cavity filling. In design of the venting cross sections, it is advisable to observe the following recommendations:

– To place the venting channels so that the premature blocking is prevented.
– Flowing in the venting cross section should be turbulent.
– Lower number of channels with large cross section have higher flow rate contrary to more channels with identical area.

- Inevitable enlargement of the venting cross section by 25% with the application of sprays.
- Channels must be sufficiently low to prevent melt from permeating into the channel under the influence of resistance pressure.
- Resistance of channels against air and gases from the mould cavity should be as low as possible.

The channels with depth within the range from 0.08 up to 0.12 mm meet the given requirements [34, 35].

The width of the venting channel can be expressed by the following relation:

$$b_k = \frac{f_k}{h_k} \tag{1.29}$$

where f_k—cross section of the channel [m^2],
h_k—depth of the channel [m],
b_k—width of the venting channels [m].

1.1.11 Casts in Die Casting

The main factors determining the possibility of production of casts by the die casting method are the volume of production, weight of cast, and its ground dimensions in the dividing plane of the mould, wall thickness, structure stiffness, requirements related to strength and tightness of the cast. The weight and dimension of the cast depend on possibilities of die casting machine, i.e. on filling chamber volume, on specific filling pressure and on locking force of the machine.

At high speed of cast cooling in the die casting mould, the regulated solidification is rather difficult to be assured. Die casting can be therefore used for casting of casts with even thickness of the wall and with specific optimal dimension which depends on type of the casting alloy. With the increase of thickness of the cast wall, the extent of the dispersed porosity increases as well, which results in reduction of the alloy density and its strength characteristics. For each alloy, there exists specific minimal thickness to be filled with the liquid metal. Such thickness also depends on physical and chemical properties of the alloy [28].

From the point of view of weight, structure, material, and dimensions, the casts can be categorized as follows:

1. According to state:

   - unrefined,
   - semi-refined,
   - unfinished,
   - machined.

2. According to the size of area projection in the dividing plane of the mould:

    - casts of up to 100 cm^2,
    - casts of up to 400 cm^2,
    - casts of up to 800 cm^2,
    - casts of up to 1600 cm^2,
    - casts of over 1600 cm^2.

3. According to complexity:

    - simple casts,
    - medium-complex casts,
    - complex casts,
    - highly complex casts.

4. According to the required properties:

    - casts with the most strict requirements—group 1,
    - casts with strict requirements—group 2,
    - casts with standard requirements—group 3,
    - undemanding casts—group 0.

1.1.11.1 Technological Properties of Casts

Technological properties of material characterize its suitability for particular processing procedure by means of which the product quality should be obtained from the point of view of its use properties. Technological properties are well observed in the die casting process when the alloy is in liquid state. Such properties include die casting temperature, gas content in the molten alloy, castability, possibility of thermal processing and weldability.

Die casting temperature is important with regard to interval of the alloy solidification, speed and time of casting due to risk of premature solidification of thin walls of the cast. Of high significance is thus amount of heat which must be transferred by the mould to cast solidification.

In fast solidification, more gases remain dissolved in the alloy than in the slow one which is important from the point of view of gas content in the molten alloy.

Castability of the used alloy considerably influences the cast quality. It depends especially on fluidity and on shrinkage of material. The fluidity is referred to as ability of material to fill the mould completely. With regard to metal mould, the inhibited shrinkage prevails in die casting. At lower temperatures, the cast can crack under the influence of shrinkage or clinks or internal stress may occur.

With regards to high porosity of casts, the thermal processing and weldability of casts is rather limited. They have been taken into consideration only lately when new die casting technologies have been developed such as vacuum die casting in case of which the casts almost without pores are produced [3, 36].

1.1.11.2 Use Properties of Casts

The use properties of casts include such properties which are influenced by the technological parameters of die casting of metals. The technological parameters primarily influencing the use properties include pressing speed, heat ratio in die casting, time of mould cavity filling and working pressure or resistance pressure. The use properties are characterized by mechanical properties, volumetric weight and strength of casts [3].

Mechanical Properties

Mechanical properties of a solid body are qualitative and quantitative expression of reaction of the body to the defined mechanical stress. Each mechanical property of material is related to respective geometry of the testing body, and analogically, the behaviour of the given material under defined conditions of mechanical stress shall depend on geometric shape and dimensions of the specific structural part. Dependences of mechanical properties on shape and dimensions are caused by discontinuous distribution of stress and deformation in the volume. Other factors to be mentioned are inhomogeneity of chemical composition and structure and anisotropy of properties which can be observed in the volume of metal materials in the metallurgical procedure of production and processing. The basic mechanical properties include the following:

- elasticity,
- plasticity,
- strength,
- toughness.

In mechanical properties of casts, the reduction of strength occurs due to porosity being formed in case of casts with thicker walls. In case of thin-walled casts, the surface layers of the walls with the value of $2a$ fill the entire cross section of the cast with thickness s (Fig. 1.23). When the wall of the cast becomes thicker, the surface layers with the value of $2a$ shall not fill the entire cross section with the thickness s (Fig. 1.24). The inner layer of cast is formed with thickness of $s1$ in case of which majority of defects of casts occur for slower solidification and mainly porosity can be observed. That causes reduction in quality and in mechanical properties of the inner layer of the casts contrary to surface layers with faster cooling and solidification [3, 37].

Volumetric Weight

With regard to inner porosity in case of casts, the weight lower than density without pores is taken into consideration at all times. In the die casting process, a specific

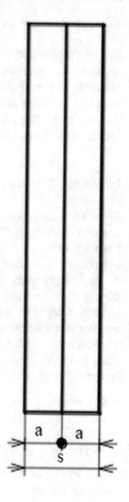

amount of gas is closed in the cast divided along the inner section into the individual
pores.

The commencement of the pressing is defined by gas volume V_0 in the metal and
with pressure p_0 and with temperature T_0. With N moles, the following equation of
state is applicable:

$$p_0 V_0 = NRT_0 \qquad (1.30)$$

where R—gas constant.

Prior to the end of pressing and under the resistance pressure, the gas volume V_1
is divided into n pores with volume V_{por} and with pressure p_1 and at temperature T_1.
The equation of state is as follows:

$$p_1 V_1 = NRT_1 \qquad (1.31)$$

Fig. 1.24 Surface layers of a wall with value of 2a shall not fill the entire cast with thickness of s; the internal layer of the cast with thickness $s1$ remains [3]

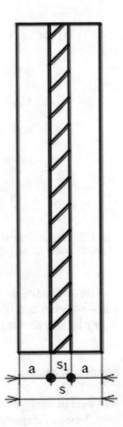

When the metal is under resistance pressure p_2, the gas pores reduce their volume to V_2. At high speed, the process can be considered to be the adiabatic one. Thus, the following relation is applicable:

$$p_1 V_1^k = p_2 V_2^k \tag{1.32}$$

where k—adiabatic gas exponent.

Volumetric weight of the alloy of the cast is defined by the following relation:

$$s = \frac{M_{\text{odl}}}{V_{\text{odl}}} \tag{1.33}$$

where M_{odl}—weight of the cast,
  V_{odl}—volume of the cast.

In die casting, the volumetric weight is influenced by speed of mould cavity filling. Except for extremely low temperatures with laminar flow the pores of bigger dimensions are formed at low speed of mould cavity filling which differs from the situation at high speed. With regard to the aforementioned fact, the volumetric weight is lower at lower speed and increases at higher speed.

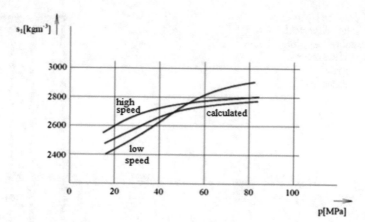

Fig. 1.25 Dependence of volume weight of aluminium alloy on resistance pressure *p* and on speed of mould cavity filling [3]

The volumetric weight of aluminium alloy was measured in dependence on resistance pressure and on high speed of mould cavity filling (Fig. 1.25). The vertical cold-chamber die casting machine was used [4, 18].

Strength of the Cast

The term of strength is referred to as resistance or fastness of material to permanent disturbance of consistence of its particles. The strength can be numerically expressed by stress in case of which the material is divided into two or more parts. The dividing is carried out by the process of disturbance under the conditions of limiting state of disturbance. The process results in fracture. Metals as well as crystalline substance can be disturbed by the following:

– cleaving (or by separating or detaching),
– shear (or by shifting or cutting).

Disturbance by shearing occurs under normal stress acting perpendicularly to the shear plane. The shear strength is expressed by normal stress. Thus, the body can be stressed by tension, bend and torsion if the fracture occurs in the plane of maximal normal stress.

Shear disturbance occurs from the shear stress which acts in the shear plane. The shear strength is expressed by shear stress. Thus, the body can be disturbed by torsion and cutting stress as well as by tension and compression if the fracture occurs in the plane of maximal normal stress [37].

Furthermore, the strength is distinguished according to means of stress as follows: strength in tension, strength in compression, bending strength, torsion strength and cutting strength. In practice and in research, the term of strength refers frequently to strength in tension. According to physical character of quantities defining the strength, the strength can be classified as follows:

- conventional or contractual strength,
- actual strength,
- ideal strength.

The conventional strength is defined by the highest conventional stress expressing specific state during straining. It is expressed as the highest conventional stress related to the initial cross section of the testing body S_0.

$$\sigma_{max} = \frac{F_{max}}{S_0} = R_m \quad [MPa] \tag{1.34}$$

The expression R_m is internationally valid and refers to conventional strength. The expression does not have any physical basis as contrary to the conventional stress the actual stress is higher in tension straining.

The actual strength is defined by the actual tension at the moment of body disturbance, i.e. by tension acting upon the fracture.

$$\overline{\sigma}_{max} = \frac{F_f}{S} = R_t \quad [MPa] \tag{1.35}$$

F_f and S refer to the values of straining force and of actual cross section at the moment of disturbance. The value R_t expresses the physical and metallurgical resistance of material against formation and spreading of fracture under given straining conditions. The actual strength of metal in the elastic area of deformation stems from the theoretical value of strength of its atom structure. If the disturbance by shearing occurs in the purely elastic state, i.e. when plastic deformation is absolutely excluded, the value of actual strength shall equal to ideal fracture strength.

The ideal strength is expressed as maximal theoretical value of strength which can be reached in case of the given material and under specific straining conditions. The solids achieve the ideal strength in the most stable state, i.e. in the shape of monocrystals without lattice damages. In case of each material, the ideal strength can be theoretically calculated from the interatomic bonding forces. As per Hooke's law, the intensity of elastic deformation depends on intensity of tension acting upon the material [37, 38].

1.1.11.3 Cast Defects

The defect is defined as the state of cast not allowing application of its use properties. From the point of view of the cast quality, the each deviation from the properties prescribed by technical standards is referred to as a defect. The main factors influencing the occurrence of defects include the structure of the die casting mould with the gating system, the venting and cooling system of the mould, the die casting machine, type of the casting alloy and its metallurgical processing, technological parameters of casting and the operation of the die casting machine. The individual factors influence each other, and reciprocal bonds exist among them. According to the technological conditions, the defect can be acceptable, non-acceptable, rectifiable or removable [3].

According to STN 42 1240 standard, the defects are classified as follows:

– defects of shape, dimensions and weight,
– surface defects,
– interruption of continuity,
– cavities,
– macroscopic inclusions and defects of macrostructure,
– defects of macrostructure,
– defects of chemical composition, incorrect physical or mechanical properties.

Short-Run Casts

In case of short-run casts, complete filling of some parts of the cast is absent. Those are usually the spots situated furthest from the ingate or the insufficiently vented spots or spots which have been fouled by grease residues.

Weld Line

It is a case of surface dents with rounded edges. A weld line occurs due to encountering of prematurely solidified metal jets. Insufficient machine pressure or insufficient metal temperature and uneven solidification are causes of the weld line formation.

Porosity

It is a case of porosity occurring due to shrinking of liquid metal in the spots of clusters and heat nodes. The porosity influences the tightness of the casts and can be detected by the X-ray tests. To remove the porosity, the number and size of heat nodes must be reduced.

Bubbles

The bubbles are characterized by small cavities with smooth surface. They are formed by the air of the filling chamber. In the process of mould cavity filling, the residual air remains in the cast when the filling is completed. The defect can be removed by gradual continuous mould filling and by effective venting.

Blisters

The defect occurs by closure of the air right below the surface of the cast wall. The main causes of blister formation include insufficient mould venting, excessive mould temperature, high temperature of alloy or low resistance pressure.

Cold Laps

The defect is caused by the encountering of prematurely solidified metal jets. The cause can be insufficient pressure of the machine or insufficient temperature of metal casting or uneven solidification of cast in the mould. To eliminate the defect, a correct setting of the technological parameters of casting must be observed.

Cracks
A crack can be defined as a surface splitting of the cast wall under cold conditions. The reason rests in strain induced by shrinking, unsuitable structure of the cast causing tension in the cast and premature removal of the cast from the mould [3, 28].

1.1.12   *Properties of Alloys at Crystallization Interval*

The liquids which follow Newton's Law (Fig. 1.26) are referred to as the Newtonian ones. The liquids not falling into the aforementioned category are the non-Newtonian ones that include Bingham liquids observing the Bingham's Law (Fig. 1.27) [3]. The liquids observing Newton's Law the following is applicable:

$$\tau = \eta \frac{dv}{dx} \tag{1.36}$$

Fig. 1.26 Scheme of flowing layers of liquid as per Newton's Law

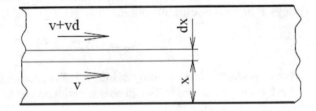

Fig. 1.27 Liquid flowing as per Bingham's Law [3]

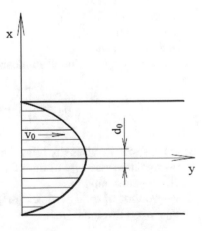

where τ—tangential stress of the layer of the flowing liquid in dependence on distance x from the edge [Pa]

η—dynamic viscosity of the flowing liquid [Nm^{-2} s]

v—speed of the layer of the flowing liquid in distance x from the edge [m s^{-1}]

x—distance from the edge perpendicular to the direction of the liquid flowing [m].

The liquids observing Bingham's Law the following is applicable:

$$\tau = \eta_p \frac{dv}{dx} + \tau_0 \qquad (1.37)$$

where τ—tangential stress of the layer of the flowing liquid in dependence on distance x from the edge [Pa]

η_p—plastic viscosity [Nm^{-2} s]

τ_0—initial stress which must be surmounted to achieve the liquid [Pa].

Binghamian liquids are alloys at interval of crystallization and suspension, for instance mould treatment grease in case of which solid particles occur in the liquid dissolvent. Not before liberation of crystallic germs closely above the liquid and crystals of the solid phase between the liquid and the solid at the interval of crystallization, the alloys observe Newton's Law. In case of mixture of solid particles floating in the liquid phase, the alloys observe Bingham's Law [39, 40].

At interval of crystallization, viscosity and initial stress should be mentioned as two basic properties. The following relation is applicable for the initial stress of alloys at interval of crystallization:

$$\tau_0 = c_1 \Delta T^a \cdot k^{b \Delta T d} \qquad (1.38)$$

where τ_0—initial stress of alloy at interval of crystallization [Pa],

ΔT—temperature reduction in contrast with temperature at the beginning of crystallic germ formation [°C],

c_1—constant [Pa °C^{-a}],

d—constant [Pa °C^{-d}],

k—constant,

a, d—exponent.

The following relation is applicable for viscosity at interval of crystallization:

$$\eta_p = \eta \left[1 + c_1 \left(\frac{V}{V_0} \right) + c_2 \left(\frac{V}{V_0} \right)^2 \right] \qquad (1.39)$$

where η_p—plastic viscosity [Nm^{-2} s],

η—dynamic viscosity of liquid alloy [Nm^{-2} s],

c_1, c_2—constants,

V—volume of solid particles [m^3],

V_0—overall volume [m^3].

1.1.13 Aluminium Alloys for Die Casting

Alloys of light non-ferrous metals (Al, Mg and Ti) are markedly applied as structural material in production of aviation and of other transport technology or of other structural components in case of which low weight is required. The general categorization of aluminium alloys (Fig. 1.28) is as follows:

1. From the point of view of increase of strength properties by thermal processing, they are classified as follows:

   – age-unhardenable,
   – age-hardenable.

2. From the point of view of processing technology, they are classified as follows:

   – alloys intended for moulding,
   – alloys intended for casting.

   Classification of aluminium alloys intended for casting is as follows [18]:

– pure aluminium,
– alloys of *Al–Si*,

   (a) eutectic alloys,
   (b) subeutectic alloys,
   (c) alloys of *Al–Si* with added magnesium,

– alloys of *Al–Si–Cu*,
– alloys of *Al–Mg*.

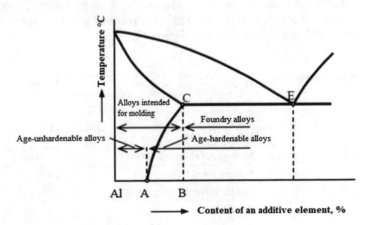

Fig. 1.28 Scheme of aluminium alloy classification [7]

1.1.13.1 Aluminium

Pure aluminium is not suitable for casting as it can be cast under the pressure with purity of 99.8%. The aluminium is used for production of casts intended for special purposes mainly in the sphere of electrical engineering for sliding rotors of electric motors. Technical aluminium is used for structural purposes (min. 99% of Al). According to the level of purity, the technical aluminium is classified as follows:

- high-conductivity aluminium—of purity Al 99.85 (maximal content of impurities reaches the value of 0.15%),
- aluminium for chemical and food industries—of purity Al 99.8 (maximal content of impurities reaches the value of 0.2%),
- aluminium for special purposes—of purity Al 99.8 (maximal content of impurities reaches the value 0.25%),
- aluminium of standard quality—of purity Al 99.5 (maximal content of impurities reaches the value 0.5%).

Owing to its extraordinary properties, aluminium disposes of wide range of applications (Table 1.3). With the given purity, aluminium is used especially for applications in case of which the combination of corrosion resistance and formability is required, for instance in production of wrappings and thin foils for food industry. Aluminium is also used in chemical and automotive industries in production of bodies of cistern trucks or of lorries, in engineering industry, for furniture, recreation and sport purposes [41].

Casting Aluminium Alloys

The group of alloys includes especially the alloys of Al–Si, Al–Cu and Al–Mg type. Contrary to the formed alloys, they contain higher amount of additives which assures rather good foundry properties. From the point of structure, the alloys are heterogeneous with the presence of eutecticum [28].

Table 1.3 Basic properties of aluminium [7]

Density [kg m^{-3}]	2700
Strength [MPa]	80
Tensibility [%]	40
Hardness [HB]	20
Modulus of elasticity [MPa]	72,000
Melting point [°C]	658
Electrical conductivity [m Ω^{-1}]	34
Formability under heat	Very good
Formability under cold conditions	Very good
Castability	Limited
Weldability	Good
Corrosion resistance	Very good

Casting aluminium alloys dispose of a range of advantages in comparison with other foundry alloys:

- good castability which considerably increases with the increase of ratio of the respective eutecticum according to the chemical composition,
- low melting temperature,
- short interval of solidification,
- content of hydrogen in the cast, which is the sole dissoluble gas in aluminium, can be minimized under adequate technological conditions,
- good chemical stability (corrosion resistance),
- good surface properties of casts,
- majority of casts dispose of low susceptibility to formation of cracks under heat.

1.1.13.2 Al–Si Alloys

A remarkable group of aluminium alloys are Al–Si casting alloys (silumins) which are applicable especially in automotive and aviation industry. Nowadays, some of the automotive components are cast exclusively from these aluminium alloys (for instance, cylinder heads for passenger vehicles, engine blocks, plungers, connecting bars, coolers, gearboxes, etc.).

Al–Si alloys crystallize according to equilibrium diagram (Fig. 1.29) of the eutectic type with limited dissolubility of silicium in aluminium. Eutectic reaction ($L \rightarrow \alpha + \beta$) is performed at temperature of 577 °C and with Si content in eutecticum ranging from 11.3 up to 12.6%.

The α phase is substitution solid solution of Si in Al with maximal dissolubility of 1.65% of Si at eutectic temperature of 577 °C and of 0.05–0.1% of Si at temperature of 200 °C. The eutecticum (E) represents in binary systems a mix of substitution

Fig. 1.29 Equilibrium diagram Al–Si [7]

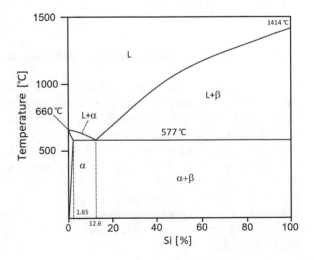

solid–liquid solution α. The mix originates directly from the melt during eutectic change.

According to the binary diagram, the Al–Si alloys can be divided on the basis of structure as follows:

- subeutectic with the structure $(\alpha + E)$ below 12.6% of Si,
- eutectic with the structure (E) of about 12.6% of Si,
- supereutectic with the structure $(E + \text{Si})$ above 12.6% of Si.

The structure and properties of foundry alloys can be in general influenced by liquid metal treatment. The treatment rests in using of a negligible amount of appropriately selected substance which influences the process of crystallization.

The treatment includes the following:

- *inoculation* which preferably influences the number of crystallic germs and which results in finer structure,
- *modification* which influences the way the crystallic germ grows and which results in morphological changes of the excluded phases.

In case of Al–Si alloys, the inoculation is employed to fine a metal matrix (of the α-phase) down and the treatment is used to achieve an optimal shape of eutectic silicium. In untreated Al–Si alloys, the eutectic silicium has a shape of hexagonal plate-like formations which in the plane of metallographic specimen appear to be dark grey needles oriented in diverse direction and placed in the light matrix of the α-phase. Such shape of eutectic silicium considerably reduces mechanical properties of silumins. Therefore, the treatment of the alloys is performed causing significant change of the structure and exclusion of eutectic silicium in a shape of sticks which in the plane of metallographic specimen appear to be rounded grains. Mechanical properties of the treated alloy considerably increase. The treatment rests in adding a negligible amount of treatment agent (master alloys) into the melt such as Na, Sr, Sb. In the aforementioned case, significant change of chemical composition of the melt must be avoided [7, 28].

1.1.13.3   Additive Elements and Their Influence on Al–Si Alloys

Additive elements occurring in Al–Si alloys (except for Si) dispose of limited ability to dissolve in aluminium and form a substitution solid solution, and at sufficiently high temperature, the elements are completely dissoluble in liquid aluminium. Yet with decreasing temperature, the dissolubility of the elements in aluminium is reduced. Ratios of additive elements which remain undissolved form inherent structural phases which are referred to as heterogeneous components of the structure. Those are frequently hard and fragile intermetallic phases deciding on physical, chemical, mechanical and technological properties of the alloy. In general, it can be assumed that these intermetallic alloys negatively influence the properties of Al–Si alloys. The extent of such negative impact depends on their size, amount, distribution and morphology.

The following are the main additives in Al–Si alloys for casts:

Copper—it allows increase of strength by hardening in the Al_2Cu phase which is contrary to the Mg_2Si more efficient. Yet it reduces corrosion resistance; therefore, it is unacceptable in alloys in food industry. During solidification, copper considerably extends the interval of silumin solidification, and thus, it can support occurrence of cracks under heat. However, copper significantly improves machinability.

Magnesium—it is added in the amount within the range from 0.3 up to 0.75%. Magnesium allows increase of strength properties by hardening in the Mg_2Si phase. With the increasing amount of Mg in alloy, the strength increases as well. Magnesium reduces formability, yet it does not reduce corrosion resistance. When in the cast state, magnesium forms intermetallic phase Mg_2Si which produces eutecticum of Al–Si–Mg_2Si with temperature of solidification of approximately 555 °C in the shape of tiny skeletal formations.

Manganese—it increases strength and corrosion resistance. At the same time, it fines the grain down and compensates negative effect of iron. Under the influence of manganese, the iron is deliberated in the form of more compact alloys of the $Al_{15}(MnFe)_3Si_2$ type in the skeletal shape or in the shape of the so-called Chinese writing. Content of manganese in Al–Si alloys should equal to approximately ½ of iron content.

Chromium, cobalt, molybdenum, nickel—they increase heat strength and eliminate harmful effect of iron; i.e., they contribute to transformation of undesired needle ferric phase Al_5FeSi to less harmful phase $Al_{15}(MnFe)_3Si_2$.

Strontium—it ranks among treatment agents. Amount of Sr depends on silicium content. It usually ranges from 0.008 up to 0.04%. If its optimal amount is exceeded, the fragile phases $AlSr_2Si_2$ occur in segregation area and porosity of casts thus increases.

Titanium, boron—both act like crystallic germs and fine the structure down—formation of fine intermetallic particles TiB_2 during crystallization. Especially in case of AlSi alloys with content of over 0.1% of Ti, the intermetallic phase $Ti(AlSi)_3$ is formed which is rather difficult to be dissolved again. With the increasing Ti content, the stability of the phase increases as well.

Lithium—it is a metal with very low density (less than $1\ g\ cm^{-3}$—unlike the water it is lighter) and with extraordinarily high reactivity. It is used in the amount of up to 5% for the purpose of reduction of density of alloys. The alloys are employed in production of components in aviation industry and in astronautics. Metallurgical and technological problems are extremely big, and scope of their utilization is thus rather limited.

Iron—it is a common impurity in aluminium alloys. In dependence on quality of base raw materials, primary aluminium contains from 0.03 up to 0.15% of Fe (on average, it ranges from 0.07 up to 0.1%). Iron negatively influences both strength and plastic properties and at the same reduces corrosion resistance. In the amount of 0.3 up to 0.5%, the iron prevents adherence (on-welding) of casts to the metal moulds (yet the iron is intentionally added into die casting alloys) and increases strength and fluidity and in higher amounts the iron increases heat strength as well. In case of contents reaching the value from 0.3 up to 0.5%, it forms undesired hard and fragile intermetallic phases (for instance Al_5FeSi and $Al_{15}(FeMn)_3Si_2$).

The iron is generally assumed to be a harmful element which generates hard and fragile intermetallic alloys. Length and density of the Fe phases increases with the increasing % of Fe which causes decrease in tensibility of casts. Increase of the Fe amount results in increase of dimensions of defects and of porosity of casts which can be demonstrated by reduction of tensibility. The ferric phases are hardly dissoluble during homogenization and deteriorate mechanical properties of alloys (strength, tensibility). At the same time, they can cause increase of tension, they have low tensibility, and they are brittle at standard ambient temperature. Despite the aforementioned, the ferric phases show extreme strength and resistance to leaking at increased temperatures [28, 42].

1.1.14 Returnable Material

Returnable material consists of gates, risers, residues from the cold chamber, rejects and other pouring admixtures in die casting. When calculating the ratio of returnable material in the batch, it is inevitable to take into account the material loss caused by metal losses during melting, by spraying in casting, swarf waste in cleaning of the cast and cast rejects. The returnable material must be degreased from the residues of greasing and cutting agents of oil origin.

The returnable material of the second melting is represented by the residues after finishing of the casts having been cast from material of the first melting. The returnable material from the third melting is represented by residues after finishing of the casts having been cast from material of the second melting.

In production of the casts in die casting of metals, the liquid alloy of lower quality is recommended from the point of view of quality and utilization of the returnable material of the second and of the third melting:

– In die casting with the use of returnable material of the second melting, its maximal content must not exceed 40%.
– In die casting with the use of returnable material of the third melting, its maximal content must not exceed 30%.
– In die casting with the use of combination of returnable material of the second and of the third melting, its maximal content must not exceed 30%.
– In die casting with the use of returnable material of the undefined melting, its maximal content must not exceed 20%.

On the basis of practical experience, it can be assumed that ratio of returnable material according to the size of the cast is from 20% in case of big casts and of up to 75% from the weight of the unfinished cast in case of small casts. When the cold-chamber machines are used, the ratio reaches 65% of the weight of the liquid metal, and in case of hot-chamber machine utilization, the value amounts to 50%.

The returnable material must be thoroughly sorted. The rejects and gating systems are considered to be clean returnable material. The impurities include residue from the filling chambers, sprayed alloy from the floor and rejects impurified by oil [3, 4, 28].

1.2 Analysis of the Production Process

Analysis of the particular production process of the die casting at the selected workplace is realized on the basis of the following:

- selection of a die casting machine,
- total of technical parameters of the selected production equipment.

1.3 Characteristics of the Cast—Testing Sample

- Analysis of chemical composition of the alloy used for production of the casts according to STN EN 1706 (STN 424310) standard under laboratory conditions with the utilization of a spectrometer (SPECTROCAST),
- preparation of testing samples.

1.4 Characteristics of Technological Parameters

Technological parameters of die casting which have been selected and shall be modified in the production process are pressing speed and pressure or resistance pressure.

1.4.1 Pressing Speed

Speed of the melt in the mould cavity depends on several factors such as speed of plunger in the filling chamber, hydrodynamic losses, ratio between the cross-sectional area and the mould half, mould venting, etc. Plunger speed ranks among the principal technological parameters influencing the cast quality. Within the framework of the research inevitable is to select experimental values of the melt flowing speed to achieve diverse homogeneity so that different modes of the mould filling (laminar, turbulent, disperse) are assured.

- Determination of experimental speed of the plunger within the range from 3.5 up to 5.5 m s^{-1} in dependence on the machine parameters.

1.4.2 Working Pressure

The pressure acting upon liquid metal in the filling chamber should assure filling of the mould at suitable speed and during optimal period of time. Hydrostatic pressure must be of sufficient intensity to surmount the resistance of solidifying metal mass in the thin cross sections of the mould cavity as well as resistance of gases remaining in the cast. The pressure is transferred from the plunger through the sprue. Transfer of hydrostatic pressure into the mould cavity is referred to as resistance pressure. To achieve the highest quality of the cast, the delay between the mould cavity filling and reaching the maximal pressure must be the shortest. Those are tenths of milliseconds at time when the cast has not been solidified. The accuracy of the cast moulding and surface roughness depend on kinetic energy of the jet.

– Determination of variants of pressure or of resistance pressure in the filling chamber.

1.4.3 Temperature of the Melt and of the Mould

Other main technological parameter to be mentioned is the temperature of the casting alloy. The alloy temperature influences especially the quality of the inner structure of the cast. The values of the melt and of the mould temperatures shall be constant in the examination process.

1.5 Cast Testing

Selection and analysis of specific strength and use properties of the casts in case of which the influence of change of technological parameters of die casting shall be examined. The selected strength and use properties of the die castings are tensile strength and porosity.

1.5.1 Examined Strength Properties of the Casts

Tensile Strength R_m

The research of properties of the materials is inevitable to be performed for their utilization under operation conditions. It is important to know the structure and properties of the produced components as well as their behaviour in case of diverse forms of stress. To determine mechanical properties, the pulley test performed with the testing samples is used.

Method:

– static pulley test,
– determination of tensile strength R_m.

Procedure:

– selection and characteristics of testing equipment,
– preparation of testing samples,
– performance of measurements according to STN EN 10002-1, EN ISO 6892 standards.

Testing equipment:

Universal testing machine TIRAtest 28100 (Fig. 1.30).

Fig. 1.30 Universal testing
machine TIRAtest 28100
[43]

1.5.2   *Examined Use Properties of the Casts*

Porosity of the Cast

The following gases are the most harmful for the aluminium alloys: water vapour, hydrogen and oxygen. Dissolubility of hydrogen is in liquid aluminium rather high and increases with the temperature. In the course of crystallization, the hydrogen is liberated in the form of pores and bubbles. At the same time, the water vapour causes formation of hydrogen bubbles. With regard to modification of the alloy temperature, the examined porosity and homogeneity of the cast shall be increasing, which is directly influenced by the temperature change.

– examined characteristics of the cast porosity:

Method:

– optical microscopy (macroscopy),

Procedure:

– selection and characteristics of testing equipment,
– selection of a set of samples from the previous testing in case of which the influence of change of mechanical properties was examined with regard to the changing technological parameters,

Fig. 1.31 Microscope 2303 Intraco Micro

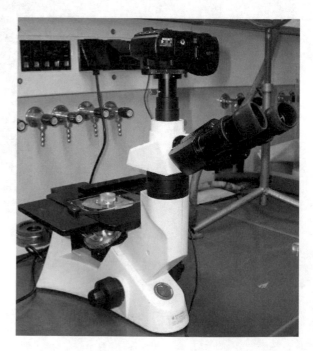

– preparation of samples (mechanical grinding and final polishing with a diamond paste),
– performance of measurements.

Equipment:
Microscope 2303 Intraco Micro

The microscope (Fig. 1.31) is designed for the analysis of porosity of metallographic specimen from the taken samples.

1.6 Characteristics of a Production Process

To perform the experiments, the die casting machine CLV 250 as shown in Fig. 1.32 was selected and placed in the operation plant of the company for production of the casts.

Basic operating parameters of the production equipment designed for production of the examined casts in the die casting process are presented in Table 1.4.

Fig. 1.32 Die casting machine CLV 250

Table 1.4 Operating parameters of production equipment

Parameters of die casting machine CLV 250	
Dimensions	2 × 3.2 × 8.7 m
Weight	27 t
Engine performance	37 kW
Locking force	600 t
Injection force	65 t
Ejection force	35 t
Min./max. mould height	400–900 mm
Max. weight of Al–Si alloy	12 kg

1.7 Characteristics of the Cast

In realization of the experiment, the AlSi9Cu3 alloy was used the chemical composition of which according to DIN EN 1706 standard is presented in Table 1.5.

It is a case of subeutectic alloy of the Al–Si–Cu type which is dominant in automotive industry. The alloy is used in die casting for production of the casts such as engine blocks and gearboxes. The presence of Cu not only does improve the machinability, but at the same time it allows spontaneous hardening after rapid cooling of casts in the water. By rapid cooling of the cast, the oversaturated solid solution α (Al) is formed which consequently disintegrates. Thus, the precipitate is generated by means of which the structure is reinforced. Hardening can be performed in case of the alloys showing dissolubility of the additive elements in the solid solution and contain the hardening phases which are liberated from the solid solution. The AlSi9Cu3 (Fe) alloy also contains much of Mg which consists of hardening phases. Thus, the cast reaches higher values of mechanical properties. This process is applied in mass production where cooling of castings in water is advantageous for easier separation of the gating system from the casting. The aforementioned procedure is consequently followed by natural hardening being referred to as ageing, i.e. a process significantly influencing further processing of the cast. To create the optimal conditions for machining, the time lapse must be observed between die casting and machining. In machining of the cast in the cast state, the material adheres onto a tool by means of which its service life shortens markedly. If the machining is performed with particular time lapse, the machining conditions shall considerably improve.

Tables 1.6 and 1.7 present physical and mechanical properties of the cast used in performance of the experiments. Figure 1.33 shows a particular cast produced in die casting. The components are utilized in the automotive industry.

Table 1.5 Chemical composition of the AlSi9Cu3 alloy according to DIN EN 1706 standard

Chemical composition in weight %

Al	Si	Cu	Mg	Mn	Fe	Zn	Ni	Sn	Cr	Ti	Pb
Residue	8.5–10.0	2.0–3.5	0.1–0.5	0.1–0.4	Max. 1	Max. 0.3	Max. 0.3	Max. 0.1	Max. 0.05	Max. 0.15	Max. 0.2

Table 1.6 Physical properties of the cast

Density [g cm^{-3}]	2700
Solid temperature [°C]	525
Liquid temperature [°C]	610

Table 1.7 Mechanical properties of the cast

Tensile strength R_m	Min. 110 MPa
Yield point Rp0.2	Min. 140 MPa
Extensibility A5	Min. <1
Hardness as per Brinell HB	Min. 80

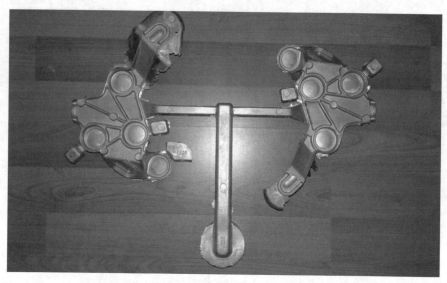

Fig. 1.33 A gating system along with cast AlSi9Cu3 produced in die casting

1.8   Technological Parameters of Die Casting

Conditions of casting expressed by technological parameters of die casting with the specific production equipment are presented in Table 1.8.

Two technological parameters of die casting with the highest influence upon final quality of the casts were selected for the research—plunger speed in mould cavity filling and pressure or resistance pressure in the mould cavity.

Experimentally selected plunger speed in mould cavity filling is as follows:

v_1—3.5 m s^{-1}

v_2—4.5 m s^{-1} (original operating value)

v_3—5.5 m s^{-1}.

Table 1.8 Technological parameters of die casting

Cast	AlSi9Cu3
Alloy temperature [°C]	660
1. Plunger speed [m/s]	0.15
2. Plunger speed [m/s]	4.5
Resistance pressure [MPa]	28
Solidification [s]	30
1. Trajectory [mm]	150
2. Trajectory [mm]	490
3. Trajectory [mm]	410

1. Speed—plunger speed in filling of the die casting chamber and of the gating system
2. Speed—plunger speed from the moment of completion of the die casting chamber filling when filling of the shaping mould cavity commences
Resistance pressure—expresses additional pressing of the plunger (multiplication) when the plunger forces into the filled mould at the end of the casting
First–third trajectories—sections of the plunger movements. Trajectory 1 represents the distance in case of which the pouring hole is locked by the plunger
Period of solidifications means the time from the mould filling up to its opening

Experimentally selected values of plunger resistance pressure in the mould cavity:
p_1—22 MPa
p_2—28 MPa (original operating value)
p_3—32 MPa.

1.9 Assessment of Tensile Strength R_m

A universal testing machine TIRAtest 28100 was used for assessment of tensile strength. The testing machine is employed in realization of pulley and compression tests. The machine consists of a stationary frame with a tensometer situated in an upper part (a device evaluating the actual loading (force)). A stationary jaw with the tensometer grips the testing body from one end, and the other end is gripped by a movable jaw fixed to a cross arm. When the cross arm starts moving, gradual loading and deformation of the testing body commences as well. Deformation of the body (its elongation) can be measured by means of an extensometer.

Tensile strength R_m ranks among the basic quantities of assessment of mechanical properties of the casts. Measurement of the quantity was performed with the testing bars (Fig. 1.34) produced from the guide channels of the gating system. The final tensile strength of the individual testing samples was assessed in dependence on the change of plunger speed in mould cavity filling. The pressing speed values were selected on the levels of 3.5, 4.5 and 5.5 m s^{-1}.

The measured values of tensile strength in dependence on change of the plunger pressing speed can be seen in Table 1.9. The measured values and the graph in

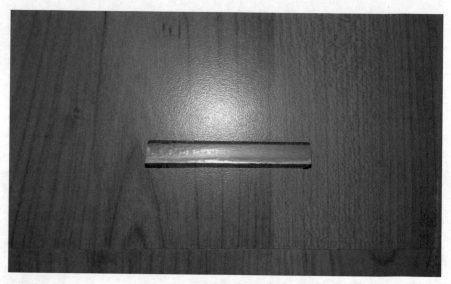

Fig. 1.34 Testing sample used in assessment of tensile strength R_m

Table 1.9 Values of tensile strength R_m in dependence on pressing speed change

Sample No.	Plunger pressing speed v_p [m s^{-1}]	Resistance pressure [MPa]	Tensile strength R_m [MPa]	Average value of strength R_{mP} [MPa]
1.1	3.5	28	175	176.3
1.2			184	
1.3			170	
2.1	4.5		159	149
2.2			147	
2.3			141	
3.1	5.5		115	119.6
3.2			123	
3.3			121	

Fig. 1.35 prove decreasing tendency of tensile strength with regard to increase of pressing speed. At higher speed values, the mould got slightly opened which caused consequent injection in the dividing plane during forcing-in. In dependence on diverse speed values of the mould cavity filling, a variety of filling types could occur—laminar, disperse, turbulent or their combination.

The values of resistance pressure in the mould cavity were determined on the levels of 22, 28 and 32 MPa. In dependence on these values, the tensile strength was measured at constant speed of plunger in the filling chamber. The values of the measured tensile strength are presented in Table 1.10. Figure 1.36 shows development

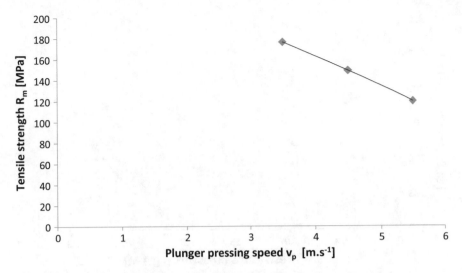

Fig. 1.35 Graphical dependence of tensile strength R_{m} on change of pressing speed

Table 1.10 Values of tensile strength R_{m} in dependence on resistance pressure change

Sample No.	Resistance pressure [MPa]	Plunger pressing speed v_{p} [m s^{-1}]	Tensile strength R_{m} [MPa]	Average value of strength R_{mp} [MPa]
1.1	22	4.5	189	183
1.2			182	
1.3			178	
2.1	28		190	192.3
2.2			198	
2.3			189	
3.1	32		210	204.3
3.2			206	
3.3			197	

of dependence of resistance pressure change on measured tensile strength with the graph showing the increase of tensile strength values along with the increase of resistance pressure in the mould cavity.

Figure 1.37 shows the comparison of tensile strength values and change of pressing speed of the plunger and of the resistance pressure.

On the basis of comparison of the measured values of tensile strength in dependence on pressing speed of the plunger and of the resistance pressure, it can be assumed that the highest values were measured at lower speed values of the plunger and at higher values of resistance pressure. The lowest measured values of tensile

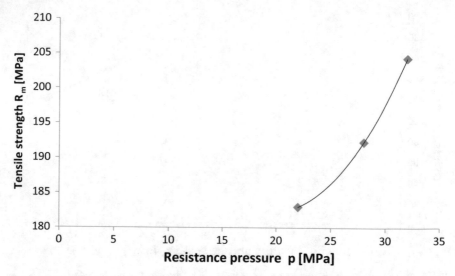

Fig. 1.36 Graphical dependence of tensile strength R_m on resistance pressure change

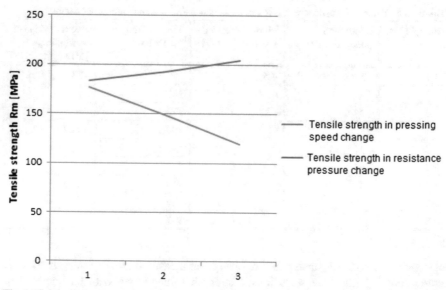

Fig. 1.37 Development of tensile strength in change of pressing speed and resistance pressure

strength could be observed at higher speed of the plunger and at lower values of the resistance pressure.

1.10   Assessment of Cast Porosity

Initially, the metallographic samples must have been prepared (Fig. 1.38) in order to assess the porosity. The ruptured testing bars used in the pulley test were employed as the basis. At the same time, the cut spots on the cast had to be selected. The samples were cut by a slitting saw MIKRON with the possibility of water cooling. The cutting parameters were selected with regard to cutting material, and compressive force acting upon a slitting disc was set to the lowest value of the loading scale. On the other hand, the revolutions were set to higher values, i.e. 3000 rev/min. The precautions were taken to obtain the cutting of the highest quality. Consequent operations were performed in a laboratory where the samples were cast into dentacryl (casting resin). Porosity assessment and microscopic observation required thorough preparation of the samples; i.e., the samples had to be ground under water with a grinding disc and then polished to achieve desired quality. The procedure was performed in the laboratory with a semi-automatic polisher Struers LaboPol-5 as shown in Fig. 1.39.

Grinding Parameters

- Grain of sand paper is of 1200.
- Revolutions of a disc are of 300 rev min^{-1}.
- Compressive force acting upon samples selected with regard to unevenness of the surface, grinding period dependent on surface quality, a single cycle lasted for 3 min, majority of samples was sufficiently ground after 2 cycles (2 × 3 min), a few samples had to be ground for a longer period.

The ground samples had to be polished further on with the use of the same machine, yet instead of the grinding disc, the diamond pastes with lubricating component

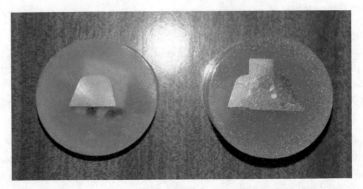

Fig. 1.38 Samples intended for analysis of porosity by means of optical microscopy

Fig. 1.39 Polishing machine Struers LaboPol-5

DiaDuo were applied with grain of 3 μm and of 1 μm. First, all samples were polished using the paste with grain of 3 μm, and consequently, they were repolished using the paste with grain of 1 μm. All the pastes dispose of special polishing discs which had to be replaced in dependence on the polishing paste. The injection paste intensity was initially adjusted manually, and then in case of very polishing, the machine determined the amount polishing paste to be added in drops onto the polishing disc during polishing of the samples.

Polishing Parameters

- Revolutions of a disc are of 150 rev min^{-1}.
- Period of a single polishing cycle is of 3 min, the samples are polished 2 × 3 min per 3 μm of the paste, and in major cases, it was 1 × 3 min per 1 μm of the paste (in a few cases, it was 2 × 3 min).
- Value of compressive force was adjusted with regard to the quality of the respective cut.

The quality of the samples having been polished on the basis of the aforementioned procedure was checked by means of a microscope with a trinocular head 2303 Intraco Micro. Once the samples were examined, it was apparent which samples were completed and which had to be repolished.

Porosity was assessed by means of optical microscopy; consequently, the image was transferred to a computer to perform a pattern analysis by computer software Stream Motion by the company of Olympus (Fig. 1.40). The produced photographs were subjected to the analysis as shown in Fig. 1.41. In case of each sample, three independent spots were selected to perform measurement and assessment of porosity. These spots had to be perfectly polished as the computer could consider the creases and dents to be the pores and the measurements results would have been thus distorted. That would have been probably far higher percentage of porosity than in the actual case. Porosity in the program Stream Motion is assessed by means of graphical filters. The filters find in the assessed sample a colour which they are adjusted to. Each of the filters disposes of defined sensibility which prevents intervention into detection zone of other filters. Prior to measurement, the filters must be adjusted to a correct sensibility so that solely pores are assessed. Once the filters are adjusted, the program analyses the objects in the respective images which differ in colour from their environment and meet all of the preset limiting conditions. The results of all values relating to the samples are saved in the tables by the program.

In assessment of the inner structure of the casts, the porosity was assessed. After performance of static tests intended for assessment of tensile strength in dependence on change of pressing speed of plunger and resistance pressure, the testing bars were used in further examination of the inner structure of the cast, i.e. of porosity. Except

Fig. 1.40 Assessment of porosity by the Steam Motion program

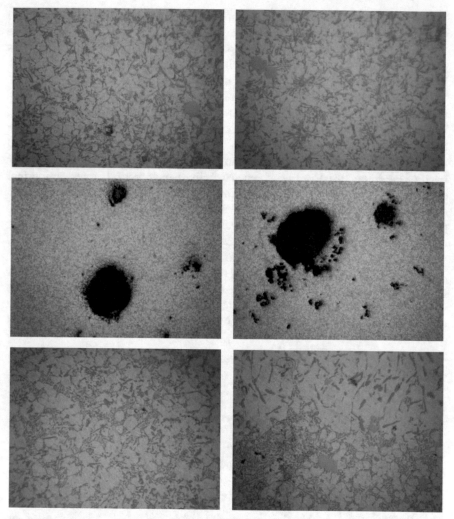

Fig. 1.41 Metallographic images of the examined samples

for that the samples were taken from other spots of the cast which were supposed to be the spots with possible occurrence of increased porosity.

The measured values of porosity of the examined samples prove the influence of the change of pressing speed of the plunger and of the resistance pressure upon inner structure of the cast (Tables 1.11 and 1.12). With the increasing pressing speed of the plunger, the increase of the values of cast porosity can be observed in Fig. 1.42. As in case of different speeds of the plunger the diverse types of the mould cavity filling occur, the porosity can be assumed to be dependent on the mode of the mould cavity filling. In assessment of the cast porosity in relation to the change of resistance

Table 1.11 Porosity values in dependence on pressing speed change of the plunger

Sample No.	Plunger pressing speed v_p [m s^{-1}]	Resistance pressure [MPa]	Porosity P [%]	Average value of porosity P_p [%]
1.1	3.5	28	2.15	2.49
1.2			2.74	
1.3			2.59	
2.1	4.5		2.89	3.29
2.2			3.78	
2.3			3.21	
3.1	5.5		4.58	4.22
3.2			4.12	
3.3			3.98	

Table 1.12 Porosity values in dependence on resistance pressure change in the mould cavity

Sample No.	Resistance pressure [MPa]	Plunger pressing speed v_p [m s^{-1}]	Porosity P [%]	Average value of porosity P_p [%]
1.1	22	4.5	3.97	4.04
1.2			4.12	
1.3			4.05	
2.1	28		3.49	3.44
2.2			3.89	
2.3			2.95	
3.1	32		2.87	2.64
3.2			2.68	
3.3			2.36	

pressure in the mould cavity, the reduction in porosity was recorded with regard to the increase of resistance pressure. The dependence is plotted in a graph in Fig. 1.43.

Figure 1.44 shows the comparison of porosity values with regard to change of the pressing speed of plunger and resistance pressure. In reciprocal comparison of values of cast porosity with the change of pressing speed of plunger and resistance pressure, the lowest value of porosity could be observed at the lowest speed of plunger and at highest value of resistance pressure in the mould cavity. The highest values of porosity were recorded at the highest speed of plunger and at the lowest value of resistance pressure in the mould cavity.

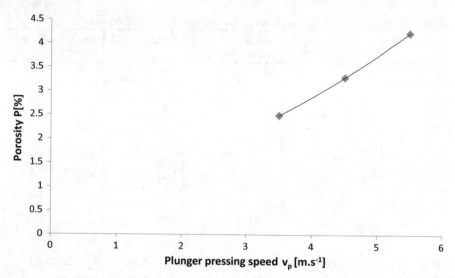

Fig. 1.42 Graphical dependence of average porosity of the cast on change of the plunger pressing speed

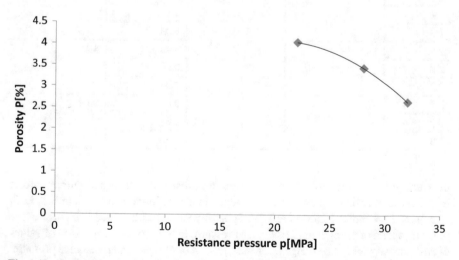

Fig. 1.43 Graphical dependence of cast porosity on resistance pressure change

1.11 Perspectives

In the die casting process, the metal is pressed at high speed into the mould cavity. The principle rests in conversion of kinetic energy into the pressure one. The entire process of die casting is influenced by a number of factors. The factors with dominant influence upon the quality of the final casts include pressing speed, specific pressure acting upon the melt and resistance pressure, period of the mould cavity filling,

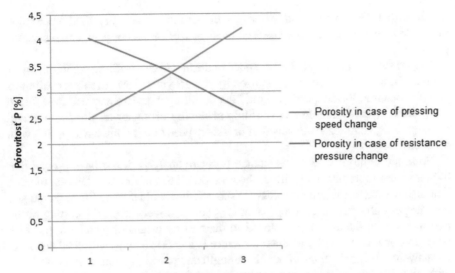

Fig. 1.44 Porosity values in change of pressing speed and resistance pressure

temperature of the casting alloy and temperature of the filling chamber and of the mould. The secondary factors influencing the entire process and cast structure, mould structure and efficiency of the die casting machine must be taken into consideration as they represent a complex of reciprocal bonds.

The measured and assessed values of the individual examined properties have proved that the selected technological parameters in a high-degree influence the final properties and quality of the casts produced in die casting. The examined technological parameter, plunger speed or speed of mould cavity filling and mode of the mould cavity filling rank among the basic factors with the highest influence upon the final quality of the cast. As a mater of course, the final properties of the casts were influenced by a number of other parameters such as melt temperature and cooling. These parameters in the form of cooling speed and mould temperature remained constant. The structural parameters were influenced solely by melt temperature which closely relates to speed of mould cavity filling. The melt speed depends on temperature as with the increase of the temperature the melt viscosity decreases which results in increase of its movability under the same pressure. Since the melt speed determines the mode of mould cavity filling, it affects the speed of melt cooling due to change of the melt volume at time of a contact with the mould surface. The aforementioned fact implies reciprocal relations among the phenomena in the mould cavity and influence upon the final properties of the cast.

Furthermore, the research has proved the connection between mechanical properties of the cast and inner structure. The measured value of tensile strength R_m and the values of porosity are interconnected in a high degree. In case of higher concentration of pores and cavities, the strength is apparently reduced, which can be shown by the results as well. With higher percentage ratio of porosity, the reduced value of tensile strength at the same speed of pressing pressure and resistance pressure was recorded. Vice versa, with lower percentage ratio of porosity, the higher value of tensile strength was recorded as well as in case of the same speed and resistance pressure.

Higher value of porosity at higher speed of the mould cavity filling points out lower degree of the mould cavity filling caused by disperse melt flow. The lower values of porosity occur at lower speed of mould cavity filling with laminar melt flow. From the point of view of mould cavity filling achieving of higher tensile strength and lower porosity requires lower speed of pressing speed or of mould cavity filling with the melt.

From the point of view of resistance pressure analysis, it has been proved that higher values more positively affect the final properties of the cast. Despite the fact that higher values of resistance pressure shorten the mould service life and elongate the idle period of the die casting machines, they positively affect the casting of the cast and reduce the air volume closed in the cast by means of which the porosity is reduced and final quality of inner structure is increased along with final strength. In analysing the final values of tensile strength and porosity, it has been proved that with higher values of resistance pressure, the value of tensile strength was higher and percentage ratio of porosity was lower. On the other hand, with lower value of resistance pressure, the value of tensile strength was lower and percentage ratio of porosity was higher. To achieve higher value of tensile strength and of lower percentage ration of porosity, it is recommended to select higher values of resistance pressure in the mould cavity.

When comparing the results of other authors having performed the research in the sphere of die casting, the influence of the examined technological parameters of die casting upon final mechanical properties proved to be similar or identical. Apart from other issues, the thesis [28] describes and examines the influence of the pressing speed upon plunger and resistance pressure upon tensile strength and porosity. When the influence of the plunger speed upon tensile strength was examined, the measure values of tensile strength were having a decreasing tendency in dependence on increase of the plunger speed. The entire process of dependence was similar to the values presented in the thesis with the exception of a case in which the considerable increase of tensile strength at the highest examined speed of the plunger was recorded. With regard to the aforementioned fact proved has been the assumption that the mechanical properties are in a high degree influenced by the mode of the mould cavity filling and the most suitable is to select lower values of the plunger speed. Concerning the comparison of the results of both theses related to influence of the resistance pressure change upon tensile strength, the values had a reducing and increasing tendency contrary to our measured values. Increase of the resistance pressure value resulted in observation of evident reduction of tensile strength and consequent increase with the highest value of resistance pressure contrary to the values of our measuring in case of which the resistance pressure increase demonstrated the increase of tensile strength. With regard to the aforementioned, the rule that the increase of resistance pressure values shall automatically result in increase of the values of tensile strength cannot be unambiguously approved. The phenomenon can be ascribed to a number of conditions which directly influence the development and result of experiments such as, for instance, different compositions and temperatures of the melt.

Conclusions similar to those referring to dependence of resistance pressure upon tensile strength can be deduced also in comparing of the values during examination of the influence of the change of plunger speed upon cast porosity. In the dissertation thesis herein the increase of porosity due to increase of plunger speed can be observed and the comparative values proved to have fluctuating development of increase and decrease during increase of pressing speed of the plunger.

When comparing the values of dependence of resistance pressure change on porosity, the identical influence was proved in both cases. During increase of the resistance pressure, the definite decrease in porosity percentage of the casts could be observed.

With regard to comparison of the researches, it can be assumed that a definite and general determination of ideal input conditions of the production process is not possible in relation to the final quality of the casts. The measured and assessed data can be in both cases considered to be the specific and valid ones for the respective combination of the selected technological parameters. The final quality and the examined qualitative properties depend on a number of criteria such as melt and mould temperature, melt composition, cast complexity and parameters of the production equipment. In assessment of the results, the conditions under which the tests were performed as well as possible occurrence of the faults must be taken into consideration.

1.12 Conclusion

The monograph deals with the research of influence of the selected technological parameters of die casting upon mechanical and use properties of aluminium casts. The thesis consists of theoretical and practical part. The theoretical part focuses on principal spheres of die casting such as basic methods of casting, the principle and characteristics, die casting machines and their categorization and description, alloys used in die casting and their characteristics and analysis of the casts and of their properties.

In realization of the experiments within the scope of practical part, the influence of the technological parameters was examined, i.e. of speed of plunger and resistance pressure in the mould cavity upon mechanical properties of the cast represented by the tensile strength R_m and percentage ratio of porosity. The results of experiments and analysis of the measured values pointed out a close relation between tensile strength and porosity. In case of the increased percentage ratio of porosity, the decrease of values of tensile strength was observed due to reduced cross section of the cast. Pressing speed of the plunger is closely connected with the mode of mould cavity filling which affected the final inner structure of the cast. The increase of resistance pressure in the mould cavity showed positive influence upon the die casting process. The consequence of higher values of resistance pressure was compression of pores during pressing which resulted in reduction of percentage ration of pores in the cast and in increase of the tensile strength values.

The research unambiguously proves direct connection and reciprocal bonds between the selected technological parameters of die casting and final mechanical and structural properties of aluminium casts. From the point of view of reaching the required quality of the casts and of increase of production efficiency and of reduction of occurrence of rejects in production, the optimal setting of technological parameters of die casting is rather significant.

In order to prevent reduction of final quality of casts desired would be to perform the research in the companies also in the future. In cases in which the companies use in production higher amount of returnable material, for instance, it is advisable to plan the checks of properties of casts at specific time intervals to avoid undesired changes of the melt composition and of the final quality of the cast. Also for the companies with increasing tendency of complaints or of defects of the produced components, it would be advisable to introduce into the production process the check focused on tests of mechanical properties and of inner structure with regard to used input technological parameters affecting the final quality of the casts. Despite the fact that the activities connected with the check and with the additional testing of the casts are costly for the companies, it is a good investment for continual increase of the quality and needs of the customers.

References

1. J. Ružbarský, E. Ragan: The stress of machine parts for pressure die casting and dies that are in contact with liquid metal—2005, in *Výrobné Inžinierstvo*, vol. 1 (2005), pp. 44–46. ISSN 1335-7972
2. A.N. Vilenneue, A.M. Samuel, F.H. Samuel, Role of trace elements in enhancing the performance of 319 aluminium foundry alloy. AFS Trans. (23) (2001)
3. Š. Michna, a kol., *Aluminum Encyclopedia* (Adin, s.r.o., Prešov, 2005)
4. E. Ragan, a kol., *Casting of Metals Under Pressure*, 1st edn. (Prešov, 2007), 392 s. ISBN 978-80-8073-979-9
5. J. Ružbarský, J. Žarnovský, Effect of input casting pressure parameters on casting structure quality, 2009, in *Quality and Reliability of Technical Systems: 14th International Scientific Conference*, Nitra, 19–20 May 2009 (SAU, Nitra, 2009), pp. 194–197. ISBN 9788055202228
6. J. Ružbarský, J. Paško, Š. Gašpár, *Techniques of Die Casting* (RAM-Verlag, Lüdenscheid, 2014), 199 pp. ISBN 978–3-942303-29-3
7. D. Bolibruchová, E. Tillová, *Foundry Alloys Al–Si* (ŽU, Žilina, 2005)
8. J. Campbell, R.A. Harding, *Introduction to Casting Technology* (The University of Birmingham, Birmingham, 1994)
9. Z. Gedeonová, *Casting Theory* (ALFA, Bratislava, 1990)
10. R. Majerník, J. Ružbarský, Methods for assessing the internal structure of the castings. 2015, in *Technofórum 2015: New Trends in Machinery and Technologies for Biosystems* (SAU, Nitra, 2015), S. 191–196. ISBN 978-80-552-1325-5
11. R. Kožený, Metallurgy and technology of aluminum foundry alloys, in *Vocational Training of Ferrous and Non-ferrous Metal Workers* (Rajecké Teplice, 2002)
12. J. Ružbarský, J. Dobránsky, J. Žarnovský, The stress of machine parts for pressure die casting and dies that are in contact with liquid metal influencing their service. 2010, in *Proceedings of the Technical University of Mining—Technical University of Ostrava—Machine Series*, vol. 56, no. 1 (2010), pp. 309–312. ISSN 1804-0993

13. L. Frommer, *Druckgiess-Technik* (Berlín, 1965)
14. P. Veles, *Mechanical Properties and Testing of Metals*, 2nd edn. (Alfa, Bratislava, 1989)
15. Ľ. Boháčik, S. Hloch, *Fluid Mechanics and Aerodynamics* (FVT TU Košice with seat in Prešove, Prešov, 2003)
16. T. Grígerová, I. Lukáč, R. Kořený, *Foundry of Non-ferrous Metals*, 1st edn. (Alfa, Bratislava, 1988)
17. E. Tillová, Alloy structure evaluation Al–Si, in *Vocational Training of Ferrous and Non-ferrous Metal Workers* (Rajecké Teplice, 2002)
18. F. Jandoš, R. Říman, A. Gemperle, *Utilization of Modern Laboratory Methods in Metallography* (STNL, Praha, 1985)
19. L. Ptáček, *Non-destructive Casting Tests* (Slévárenství, 2003)
20. J. Ružbarský, Influence of technological parameters of die casting on casting structure, in *Slévárenství*, vol. 57, no. 3–4 (2009), pp. 80–82. ISSN 0037-6825
21. M. Edwards, M. Endean, *Manufacturing with Materials* (The Open University, U.K., 1990)
22. Ľ. Boháčik, S. Hloch, *Thermomechanics* (FVT TU Košice with seat in Prešove, Prešov, 2006)
23. J. Ružbarský, J. Žarnovský, Dynamics of die casting. 2014, in *Quality, Technology, Diagnostics in Technical Systems* (SAU, Nitra, 2014), pp. 140–145. ISBN 978-80-552-1194-7
24. M.A.A. Najafabai, S. Khan, A. Ourdjini, R. Elliot, The flake fibre transition in aluminium-silicon eutectic alloys. Cast Metals **8**(1) (1994)
25. E. Ragan, J. Ružbarský, Parameters stating in the die cavity 2006, in *Materiálové inžinierstvo*, vol. 3 (2006), pp. 21–22. ISSN 1335-0803
26. P. Řezníček, Evaluation of melt quality, in *Current Knowledge of Materials and Technological Processes in Aluminum Alloy Foundry* (Mladá Boleslav, 1995)
27. M. Jermy, *Fluid Mechanics A Course Reader* (Mechanical Engineering Dept., University of Canterbury, 2005), pp. 5–10
28. V. Gorjany, P.J. Mauk, O. Mzronova, *Thixomolding Giesserei* **2**(2006)
29. O. Irretier, *Heat Treatment of Aluminum Components—A Fully Automated Design Concept Reduces Heat Treatment Costs* (Slévarenství, 2001)
30. J. Malik, *Foundry Machines and Equipment* (HF TU Košice, 2006)
31. T. Podrábsky, P. Skočovský, Colour contrast in practical metallography. Acta Metall. Slov. 8 (2002)
32. I. Pavlík, J. Chrást, *Furnaces for Melting, Maintaining and Casting Iron and Non-ferrous Metal Alloys* (Slévárenství, 1994)
33. L. Ptáček, *Defects in Aluminum Alloy Casting* (Slévárenství, 1998)
34. H.K. Barton, Giesserei-Praxis **4**, 102 (1966)
35. I. Durmis, Heat treatment of non-ferrous metals and their alloys, in *Vocational Training of Ferrous and Non-ferrous Metal Workers* (RT, 2002)
36. E.J. Vinarcik, *High Integrity Die Casting Processes* (Wiley, New York, 2003)
37. J. Ružbarský, R. Majerník, P. Hrabě, P. Valášek, Dynamics of treatment device for die casting of metals. 2016, in *Key Engineering Materials: Operation and Diagnostics of Machines and Production Systems Operational States 3*, vol. 669 (2016), pp. 327–334. ISSN 1013-9826
38. E. Ragan, J. Ružbarský, I. Andrejčák, Development and prospects of pressure die casting—2005, in *Slévárenství*, vol. 2–3 (2005), pp. 84–87. ISSN 0037-6825
39. Š. Michna, P. Lukáč, *Color Contrast, Structure and Defects in Aluminum and its Alloys* (Delta print, Děčín, 2003)
40. S. Murali, K.S. Raman, K.S. Murthy, Al–Si cast alloys: a new approach of property improvement. Trans. AFS **165** (1996)
41. R. Majerník, J. Ružbarský, Process of assessment of porosity of die-casts based on Al–Si. 2015, in *Posterus.sk*, vol. 9 (2015), pp. 1–8. ISSN 1338-0087
42. F. Paray, J.E. Gruyleski, Microstructure—mechanical property relationships in a 356 alloy. Part I: microstructure. Cast Metals **7**(1) (1994)
43. F.C. Bennet, What to expect from a vacuum die casting. Br. Foundryman **54**(2), 54–58, 507 (1961)

Printed in the United States
By Bookmasters